T0187734

# Models for Design

## Electrical Calculations for Industrial Plants

# Models for Design

## Electrical Calculations for Industrial Plants

Robert E. Henry, PE

**CRC Press**
Taylor & Francis Group
Boca Raton London New York

CRC Press is an imprint of the
Taylor & Francis Group, an **informa** business

CRC Press
Taylor & Francis Group
6000 Broken Sound Parkway NW, Suite 300
Boca Raton, FL 33487-2742

First issued in paperback 2019

© 2018 by Taylor & Francis Group, LLC
CRC Press is an imprint of Taylor & Francis Group, an Informa business

No claim to original U.S. Government works

ISBN-13: 978-1-138-50468-4 (hbk)
ISBN-13: 978-0-367-89188-6 (pbk)

This book contains information obtained from authentic and highly regarded sources. Reasonable efforts have been made to publish reliable data and information, but the author and publisher cannot assume responsibility for the validity of all materials or the consequences of their use. The authors and publishers have attempted to trace the copyright holders of all material reproduced in this publication and apologize to copyright holders if permission to publish in this form has not been obtained. If any copyright material has not been acknowledged please write and let us know so we may rectify in any future reprint.

Except as permitted under U.S. Copyright Law, no part of this book may be reprinted, reproduced, transmitted, or utilized in any form by any electronic, mechanical, or other means, now known or hereafter invented, including photocopying, microfilming, and recording, or in any information storage or retrieval system, without written permission from the publishers.

For permission to photocopy or use material electronically from this work, please access www.copyright. com (http://www.copyright.com/) or contact the Copyright Clearance Center, Inc. (CCC), 222 Rosewood Drive, Danvers, MA 01923, 978-750-8400. CCC is a not-for-profit organization that provides licenses and registration for a variety of users. For organizations that have been granted a photocopy license by the CCC, a separate system of payment has been arranged.

**Trademark Notice:** Product or corporate names may be trademarks or registered trademarks and are used only for identification and explanation without intent to infringe.

**Visit the Taylor & Francis Web site at**
**http://www.taylorandfrancis.com**

**and the CRC Press Web site at**
**http://www.crcpress.com**

*The help in editing this book by Mark E. Henry is gratefully acknowledged. The support and review by Sharon A. Henry, Atty, is deeply appreciated. Kyra Lindholm and Deepa Kalaichelvan helped in so many ways. This book is dedicated to all of them.*

# Contents

## Section II   One-Line Designs

## Section III   Fast Transient Calculations

# List of Figures

# List of Tables

# Acknowledgments

Grateful acknowledgment is made for authorization and permission to use and reprint material from the following sources:

© 2003 IEEE, Proceedings, Cable Sizing for Fast Transient Loads, R. E. Henry, 2003 Industrial and & Chemical Power Society Conference, St. Louis, MO

© 2004 IEEE, Proceedings, Response of Power Cables to Fast Transient Loads, R. E. Henry, IEEE Industry Appl. Conf. 39th Annual Meeting, Seattle, WA

© 2015 IEEE, Proceedings, Fast Transient Loads of Low Voltage Power Cables, R. E. Henry, Ind. and Commercial Power Systems 51st Technical Conf. Ind. Applications Soc., Calgary, Alberta, Canada

© 2002 IEEE, IEEE Std. 242, Recommended Practice for Protection and Coordination of Industrial and Commercial Power Systems—Buff Book, ANSI/IEEE Std. 242-86

# Author

 **Robert E. Henry, PE**, has more than 50 years of engineering experience and earned a BSEE and an MS in industrial management. He spent his early years in the aerospace industry and worked on the design of the navigation and power systems for the Apollo spacecraft. Most of his career has been in industrial plant design. Power generation, substations, and design of nuclear plants constitute most of his experience. He also has experience in food, petrochemical, mining, and paper industries. He authored a paper for the Institute of Navigation on orbital rendezvous. He has also authored papers on heat flow in electronic equipment and on power cables conducting kiloampere loads lasting several minutes. He is president of R. E. Henry P.E., LLC.

# 1

## *Introduction*

The chapters herein are intended as go-bys. Each is an example of an analysis and included calculations. The relations and formulas are all given in the references. However, cable tables are developed in Chapters 7 and 8. Nothing illustrates a relation like a calculation. Working problems at the end of a chapter is how one learns the material. With this in mind, I thought that presenting the topics herein in a traditional-text format would be a useful undertaking. I haven't included problems at the end of the chapter, but I do include examples. The level of knowledge would be for the engineer reader with an undergraduate degree not competent with a specialist's background.

There are a few relationships that I do expect the reader to know. An example is that for finding the rated current of a transformer, given the number of phases, the rated voltage, and the kVA rating. Also see Question 5 in the Fun Test at the end of this chapter. And the reader should be familiar with current transformers (CTs) and voltage transformers (VTs, and rarely PTs.) These are readily learned from the examples herein if the reader doesn't know these relationships. Also, the reader should know the inside of an oil-filled transformer, what its core looks like, and have a rough idea of what a winding looks like. The same for a motor. He must know the difference between power and energy.

A long time ago, my math professor said that when he was giving oral examinations to his master's degree candidates, he would ask them to name the fundamental theorem of differential calculus. He said that none of his candidates could ever name it. Years later, I thought I would compose a little test for newly graduated electrical engineers whom I was interviewing, who were applying for employment with my company. The test was to be comprised of a few questions like my math professor's, designed so that few could answer any of the questions even though they should know the answers. How many thousandths of an inch is the setting of a spark plug's gap was one question, as well as what is the fundamental theorem of differential calculus. The test was to make the candidate squirm and think he would not receive an offer. It was a great ice breaker. (If you are curious, the test is included at the end of this chapter, and with answers.)

It taught me that no one knows everything, and don't assume that knowledge of a topic is widely known. I know many competent design engineers who know a great deal about variable-speed drives and uninterruptable power supplies, how to model a motor, and how do motors perform at other than 50 or 60 Hz. I also know many, many more engineers who don't know much about these topics. Also, few electrical engineers know how a power cable's ampacity is arrived at, and very, very few know how a power cable responds to kiloampere current spikes lasting several minutes.

Engineering schools have not taught variable frequency effects in motors until recently. Nor did they teach the big energy savings that can be realized by not including traditional valves that control flow in industrial applications. Decades ago, I was taught analog audio circuits and their response to variable voice frequencies. This did not give me much preparation for variable frequency effects in motors. These effects are within the scopes of both electrical and mechanical engineering. Few universities teach electrical power and its applications anyway. And the development of power transistors, which made variable-speed motors practical, is only about three decades old.

I also thought I might include other topics in industrial power, such as uninterruptable power supplies. Other related topics are storage batteries and switchgear at 13.8 kV in a main-tie-main arrangement, the maximum demand that medium-voltage switchgear can supply. Three-winding transformers used as step-ups in generating plants is another favorite topic. Also, not all of the calculations are electrical. Some analyze the heat flow in a battery, cable, or motor. Here, the physics of heat flow is drawn upon. My models are usually crude. Thereby, the calculation results are approximate. However, they offer insight and are suitable for verifying computer calculations based on detailed models.

Motors are fundamental to industrial plants. They are among the few oldest electrical products. Their complication isn't widely familiar. Where their operation is required for short periods at loads beyond their rated horsepower, is covered in two chapters and Appendix B.

This book is in units of feet, pounds of force, seconds, degrees Celsius, watts, hertz, amperes, and volts. These are electrical engineering units found in most texts that were published during the last half of the last century. Also, for heat, watts and watt-hours, not Btu per hour and Btu's. Watts and watt-hours are easier and more direct with electrical heat dissipation, and to heat flow in power cables and motors. Motor units are horsepower and lb-ft of torque. The focus is also on 60 Hz and American circular mils. Most product data on the Internet is in these units. Not that the methods cannot be applied to 50 Hz and square millimeters. I leave that to the reader. Also, my circuit calculations use impedance values, not per-unit values. I believe the number crunching is the same either way. In seeing the results, I believe impedance values are closer to Ohm's law and better show the sense of the calculation. Use what is easier for you.

I have designed European industrial plants and generating stations. Once one starts working on a European project, thinking in European units is easy.

The engineering standards used in the western hemisphere are equivalent to European and Asian standards. With many products, their electrical standards are the same or used intermingled. The performance standards and test procedures for uninterruptible power supplies are common in both hemispheres, for example. The most compelling equivalence is that universal electrical units are common throughout the world and have been since the beginning of electricity as a commodity. Electricity as a tariffed commodity began in Chicago with Commonwealth Edison's AC system about 1900. And Edison, with the others, used volts and amperes as electricity evolved in the 1870s. Volts and amperes came from the scientific flowering in late-eighteenth-century Europe. An effort to drop feet, mass, and pounds of force in favor of European units continues.

In calculations, I use four significant figures. To think this improves accuracy is ridiculous. It is simply convenient for me. Three significant figures are plenty. Usually, the elements of what the calculation addresses aren't known well enough to justify more than two significant figures. Knowledge of the calculation's elements determines accuracy, not the number of significant figures. It used to be that the number of significant figures *signified* how accurately the calculation's elements were known, but this seems to have fallen away.

The topics presented in this book are listed in the Table of Contents. Start at any chapter, there is no beginning or end. Although some of the chapters are linked, each stands on its own, more or less. Computer programs for electrical analysis are not part of the topics. They have their place with batches of data and repetition of calculations. But garbage in is garbage out. There is no substitute for mastery of the engineering behind the computer programs. Independently verifying results by plain old analysis with a pencil, paper, and calculator isn't going out of style.

Fast transient loads are kiloampere loads that last for several minutes. I did a lot of original engineering work with fast transient heating of power cables and sizing them for such loads. That work is included herein. Shown herein is that sizing motor feeders by methods per the *National Electrical Code* using tables for motor current is overly conservative and needlessly costly. The methods using tables for motor currents are inappropriate for motors with varying load. (The code allows other methods applied under engineering supervision.) Nine of this book's 15 chapters address fast transient loads. Four of the nine chapters are about fast transient heating of motors.

Engineering is the art of applying science to human needs. It is a learned profession. In my case, the science was mostly physics, of the Newtonian kind. I have the honor of having helped to design the Apollo Lunar Module, which was done mostly at Bethpage, New York, by Grumman Aerospace Corp. now absorbed by Northrop Grumman. (Through performing tests

in a vacuum chamber, I also established that the accuracy of the naviga-
tion sextant in the command module was 2.3 s of arc, and not the intended
1.2 s of arc. The fix was to energize the trunnion's resolver's stator, not
its rotor.)

There is a void between the capability of a product and its application. The
product development engineer of a variable-frequency drive is intimate with
its control software but may be weak in the range of required performance
and relative costs of installations. From knowing the design of the plant, the
particular needs that the drive must meet is a separate specialty. This perspective, of requirements for transformers, cables, motors, and the like, is the
perspective of this book.

I hate errors in texts. Errors stop the reader just when she is working hard
to follow the topic in the sentence. It makes the reader angry. I have worked
hard to correct all errors I have found, but some must remain. All errors
herein are mine, and I'm sorry for them.

**Robert E. Henry, PE**

## Fun Test

### Questions

1. What is the fundamental theorem of differential calculus?
2. During factory testing, at what output kVA is a transformer's capacity measured (standard procedure)?
3. How is a transformer's impedance measured?
4. In balanced three-phase circuit analysis, how many phases are modeled at a time?
5. If analyzing a transformer in its connected circuits, if the transformer's load impedance on the secondary side is 1, what is this transformer's load impedance on the primary side if the turns ratio is 3:1?
6. What causes transformer inrush when a transformer is energized?
7. What is the setting of a spark plug's gap?
8. In music theory, what is the classical sonata form? (asked if testee said yes when asked if testee took a music appreciation course)
9. If charging a storage device with an available source of energy, can all of the available energy be coupled into the storage device?
10. What fundamental law governs this process?

**Answers on the Next Page**

## Answers

1. Given a function, y = f(x), for the value of delta y = f(delta x), as the value of delta x approaches zero, delta y approaches dy.

2. Zero output. The secondary terminals are shorted together with shunts (or using CTs) in the shorting bars to measure phase currents.

3. With the transformer so connected, the primary voltages are raised from zero to the value where secondary currents are at their rated value. The percent of measured primary voltage, at rated shorted secondary current, is the percent impedance. (It results from, of all the flux generated by the windings, not all is being induced into the core, and losses.)

4. One. It represents each of the balanced phases. Its applied voltage is the single-phase value, the rated primary voltage divided by the square root of three.

5. Nine.

6. The residual flux in the core. If the transformer is somehow turned off at the exact instant when the core is at zero flux, and if turned back on at the same voltage when turned off, there will be no inrush. This is possible with three phases, each shifted 120° from the other, if each phase has the same current–voltage relationship.

7. From 15 to 75 thousands of an inch, depending on the engine requirements and the spark plug.

8. Two themes (songs) are in a movement. Each is introduced, then developed in the middle of the movement and mixed, then restated at the end.

9. It is impossible to store all of the available transformed energy because losses are inherent. Also, losses are inherent if discharging the storage device.

10. The First Law of Thermodynamics: Energy cannot be created or destroyed but can only be transformed.

We beat on, with our models, against the stronger wind of skeptics, back into the past. Were they skeptics? Or were they just green shackled, unable to grasp change? Unable to spend the grueling labor of the mind to gain the visions that the models radiated?

# Section I

# Plant Equipment

# 2

## Using an Uninterruptible Power Supply to Feed a Variable-Speed Drive

### Introduction

This chapter presents the application of a variable-frequency drive connected to a motor driving a fan. The fan is ventilating a building with airflow that is essential to a process within the building. With loss of normal building power, the process is interrupted, but airflow must be maintained for 20 min longer during the power outage. This flow purges the process of gases that could otherwise accumulate and become explosive. However, 40% flow, reduced from full normal, is required. Because there is no backup generator, the variable-frequency drive is switched to an uninterruptible power supply having a 20 min battery.

It may be uncommon to find an uninterruptible power supply feeding a drive that supplies a fan.

However, this application is necessary

- Where power interruption cannot be tolerated
- Where a backup generator isn't available
- Where power quality is poor

Sometimes, critical loads must operate after loss of normal power and during the interval when the backup source is being started, especially when starting attempts are prolonged.

The combination of an uninterruptible power supply (UPS) supplying a variable-frequency drive first begs the question: why isn't a front end of a UPS, with its rectifier, battery, and DC bus, married to a variable-frequency drive's back end? The drive's part would have a controller, with its inverter and motor control. The answers are that the demand for such a combination is uncommon, and UPS design requirements and variable-frequency drive requirements are different. The analysis presented herein is useful in other separate applications of drives and UPSs. The examples are usable for matching UPSs to other loads besides drives. An example of a drive supplying a motor with a fast transient load is given in Chapter 10.

This discussion focuses on sizing an uninterruptible power supply (UPS) to power a variable-frequency drive (VFD), which drives a motor and its fan load. Although pumps follow the same principles as fans, addressed herein are only fans. This discussion is limited to

a. A fan forcing airflow and driven by a motor
b. A variable-frequency drive driving the motor
c. A standard 460 V three-phase 60 Hz induction motor, totally enclosed and fan cooled, with horsepower rating from 60 to 600 Hp
d. A variable-frequency drive supplied from a 480 V source
e. Both VFDs and UPSs of the low-harmonic type with active filtering

The airflow is regarded as incompressible, and the Fan Laws apply (see Bibliography). However, in some applications, a large flow head requires a turbocompressor, and then compressible flow applies. Air compressor flow is an isentropic process, with the process inefficient in the turbocompressor. Hence, the turbocompressor manufacturer provides the performance characteristics tailored to the needed flow. The manufacturer of a fan or a turbocompressor provides performance data and curves to define the most accurate information for the VFD's operation. Figure 2.1 is an example of a turbocompressor's performance curves and data.

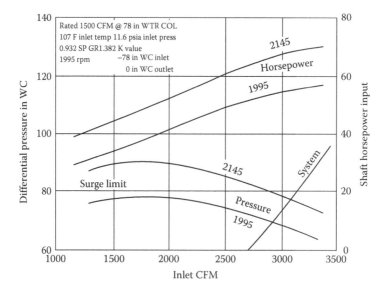

**FIGURE 2.1**
Turbocompressor performance.

## Variable-Frequency Drives

A drive has three sections, an output converter that inverts DC into the output phase currents applied to the motor, an input converter that rectifies the supplied voltage and current, and that supplies a DC link or bus. This bus, with ripple filter, supplies the inverter. The output inverter is a current source and supplies a current to the motor. The inverter shapes the current in frequency, waveform, and amplitude. Motor voltage reduces with reduced frequency. The applied motor voltage adjusts to supply the motor's complex current, supplying real power for the shaft and reactive power for the motor's excitation. A selected parameter in the drive's configuring provides this performance.

A variable-frequency drive cannot supply fault currents. A drive is comprised of six large power transistors that conduct in pulses that construct the output phase currents, in waveform and frequency. These power transistors (known as IGBTs, integrated-gate bipolar transistors) would melt if required to conduct fault currents. To protect the transistors, means are included that stop the transistors from conducting when the output circuit is faulted.

The output transistors are forced-air cooled and can overheat. The performance limits of the forced-air cooling fix the drive's rating. Compared with a motor, which is also limited by its thermal capability, a motor can overheat in minutes, while a drive can overheat in seconds.

Discussed herein are IGBTs and pulse-width modulated outputs that these devices can deliver. AC drives that utilize silicon-controlled rectifier diodes (thyristors or SCRs) in their outputs are not discussed. These types have filtering features and harmonic performances that are usually more severe than IGBT equipment.

The drive's power source would see only a rectifier and DC load when looking into the drive's input. The input power has a very high power factor that approaches unity. The harmonic content of the current that is drawn by the rectifier is actively filtered, yielding a low distortion.

The input current can be less than the output current when the input and output voltages are equal. This is because of the drive's reactive current supplied to the motor's excitation, which is generated by the inverter power unit. With the drive's high efficiency, the input current is basically resistive, while the drive's output current is resistive and reactive in supplying the motor.

Drives can utilize diodes for rectifying their input. With diodes, more elaborate DC filtering, and input filtering upstream from their diode sections, is required to control harmonic current distortion. Drives with IGBTs in both their inputs and outputs are discussed in this chapter. The input and output converters are comparable units. Both have six IGBTs. They differ in their controls. Like the output converter, the input converter cannot withstand a fault current on its DC bus. Excessive rectifier current turns off the rectifier.

Input rectifiers utilizing IGBTs actively filter the harmonic currents by measuring the harmonics in the rectified currents and adding their negatives to the input. This canceling out of the harmonics yields very low input current distortion.

The output current is complex, that is, it consists of two currents, real and reactive. The real current generates torque in the motor, and the torque drives the shaft. At any instant, the motor *passes through* to the inverter the steady-state torque. The torque generated by the inverter exceeds this torque to increase shaft speed and also to meet any added dynamic torque demanded by the load on the shaft. The inverter, in turn, loads the DC bus with these torque components as *DC power*. The rectifier sees a change in bus current demand and supplies current to regulate the DC voltage. Thus, the torque applied to the shaft, demanded by the fan, is passed through to the drive's input.

The reactive current supplies the motor's excitation. The torque driving the shaft results from the rotating flux in the motor's air gap. This flux results from the reactive current supplied by the drive. (The rotating air-gap flux consists of the exciting flux and the rotor's back emf flux.) Thus, the real and reactive currents are interrelated. *Looking into the drive by its power source, the drive's input is a dynamic resistive load.*

It is necessary to understand a drive to provide for its power supply. Its supply must meet the drive's demand for electrical power. The demand is the sum of the steady-state power and the dynamic power. The supply must be capable of providing both. The motor can deliver more than rated torque but only for dynamic loads, which aren't sustained and are too brief to overheat the motor.

Motor breakdown torque is about 200% of rated torque. The motor can operate for a few minutes with overload near this torque. The drive can also supply overload currents. Typically, a 125% overload can be supplied for half a minute, and a 150% overload can be supplied for one-sixth of a minute. Hence, in overload capability terms, the motor is stronger than the drive.

## Uninterruptible Power Supplies

The differences between a drive and a UPS are minor. Both have the same rectifier, DC bus, and inverter sections. The UPS's input section also has a battery. The output inverter operates at a fixed 60 Hz and closely regulates its voltage and frequency. The input rectifier regulates its DC voltage, supplies the inverter, and also charges its battery. Therefore, its input section is larger than its output section. Because batteries cannot tolerate ripple, the DC bus filter suppresses ripple more than in a drive. Strict input harmonic control is provided in the input converter, which is an active input harmonic filter.

Not part of a drive, the UPS also has a bypass source and switch, which transfers the load to the bypass source. The transfer switch responds to trouble in the output converter. The transfer switch is also called a static switch because it can transfer the load in about a quarter cycle. This is achieved with solid-state power devices. The static transfer switch also transfers a faulted output to the bypass source. To clear the fault, the bypass source can deliver a greater fault current than the inverter. A manual maintenance switch for bypassing the UPS is also included.

Many commercial UPSs have a limited capacity in their solid-state static switches. To limit their size, the switches have limited current capacity. Connected to a panelboard output, a fault on a feeder fed from the panelboard interrupts power to the other loads. Until its circuit breaker opens and clears the faulted feeder, the panelboard's bus voltage is near zero. A circuit breaker can open in a cycle if interrupting a strong-enough current. With limited clearing current, the circuit breaker takes much too long to open.

---

## Harmonic Distortion

Harmonic distortion standards for industrial plants limit harmonics at three points of connection: these are the UPS's supply bus, the UPS output, and the motor connections. At each connection point, the voltage waveform and the supplied and drawn currents are subject to distortion limits. Generally, all are limited to 5%.

Included in this book's other chapters are motor feeders that are about 1 km long. The motor current from the drive must not exceed 5% distortion for this application. With a long feeder, standing waves can result in motor voltage greater than rated.

The plant AC bus feeding the UPS is not usually distorted more than about 3%. The UPS operates with voltage distortion up to 5% per standards. During qualification testing, the full-load current distortion is measured. Total harmonic current distortion can be 4.4%, having some fifth with mostly third.

The drive's rectifier input current distortion depends on the weakness of its power source. The weaker the source, the greater the current distortion. With a UPS source, its inverter output is so well regulated (typically ±2%) that it appears as a very strong source. When operating from its battery, the battery voltage varies ±12.5% (2.25–1.75 V/lead–acid cell). With this variation, output voltage regulation is less than 2%, and harmonic distortion is about 2%. During factory testing of a 200 kVA UPS, voltage distortion was 0.7%. With one phase open, current distortion was 1.2%.

The drive's input is actively filtered. Harmonic currents are measured on two of the rectifier's phases and the DC bus. From these, the rectifier generates a negative of the harmonic currents and injects them into the input. The sum of these currents results in an input current distortion of less than 5%. During factory testing, a 950 kVA drive was measured with input current distortion of less than 5%.

Early UPSs and drives used diodes and thyristors for rectifying and inverting. Common harmonic distortions were higher than 20%. With use of power transistors and pulse-width modulation, harmonic distortion is no longer a problem in plant power systems.

## Torque

Torque applied to a shaft makes the shaft turn. Force equals mass times acceleration in linear terms. When a force acts on a mass over a distance, work is done. Force times distance is foot-pounds of work. In rotation, torque times circumference is foot-pounds of work.

If a force, F, acts over a circumference, work is done:

F times circumference equals ft-lb

$F (2\Pi R) =$ ft-lb in one revolution

$FR = \tau$

$2\Pi \tau =$ ft-lb in one revolution

$(2\Pi \tau)(rpm) =$ ft-lb/min

Using feet, pounds of force, and seconds

$\tau$ is torque, lb-ft

$\alpha$ is rotational acceleration, rev/s$^2$

$J$ is rotational inertia

## Rotating Inertia

A rotating mass has a characteristic rotational inertia. Rotational inertia has a characteristic radius of gyration. A bicycle wheel has its mass mainly in its rim and tire. Its radius of gyration is large for its weight. A steel cylinder of the same weight has a small radius of gyration because its weight is near its axis. Also, the more the weight out from the axis, the larger the radius of gyration. A bicycle with a heavier rim and tire is more stable than a bicycle with a lighter rim and tire. Riding with no hands on the handlebar is easier with the heavier wheel. In MG 1, the symbol for radius of gyration is k (also, K or R). The common term for a fan or pump is $Wk^2$ (lb-ft$^2$), where W is rotating weight. The manufacturer knows the value of Wk² (or WR²) for his fan or pump:

$$J = (W/g)k^2 \quad g = 32.16 \, \text{ft/s}^2$$

J is (rotational mass) (square of radius of gyration), lb-ft-s²/rev

ω is rotational speed, rev/s

t is time (sec) or $T - t_0$

$$\tau = J\alpha \, (\text{like } F = ma): \alpha = \tau/J$$

$$\alpha t = (\tau/J)t = \omega - \omega_0$$

These basic formulas are for the J torque, that is, *changes* in shaft rpm. Because Wk²/g has units of lb-ft-s²/rev, *rps* must be used, that is, rpm/60. NEMA MG 1 lists maximum values of Wk² that motors can safely accelerate. This list is Table 12-7 in MG 1. *This list is based on across-the-line starting.* For a 200 Hp, 3570 rpm motor, the value is 172 lb-ft². The compressor in Figure 2.1, having a 200 Hp motor, for example, has a value of 210.2 lb-ft². Thus, because the table value is based on across-the-line starting, the table value is a guide. Total Wk² is that of the fan plus that of the motor. Motor values of Wk² are listed in Table 2.1. See the column to the right of torques.

At a *fixed* rpm

$$(2\Pi \tau)(\text{rpm}) = \text{ft-lb/min}$$

$$1 \, \text{Hp} = 33{,}000 \, \text{ft-lb/min}$$

$$\text{Hp} = (2\Pi \tau)(\text{rpm})/(33{,}000 \, \text{ft-lb/min/Hp})$$

$$\tau = \text{Hp}(33{,}000)/2\Pi \, \text{rpm} = \text{Hp}(5{,}255/\text{rpm})$$

$$\text{kW} = \text{Hp}(0.746)$$

$$\text{kW} = (2\Pi \tau \, \text{rpm})(0.746)/33{,}000$$

$$\text{kW} = (\text{rpm})0.14197 \, 10^{-3} \, \tau$$

$$\text{kW of a drive} = (\text{rpm})0.14197 \, 10^{-3} \, \tau/\text{motor efficiency}$$

$$\text{kW of a drive} = (\text{rpm})0.14197 \, 10^{-3} \, \tau/\eta_{\text{mot}}$$

$$\text{The input kW to a drive} = (\text{rpm})0.14197 \, 10^{-3} \, \tau/\eta_{\text{mot}} \, \eta_{\text{drive}}$$

For example, a 400 Hp motor is loaded at 160 Hp when operating at 310 rpm. The motor's efficiency is 85% at this load, and the variable-speed drive's efficiency is 94%.

**TABLE 2.1**

Typical Motor Performance Data

460 V 60 Hz Design B Energy-Efficient TEFC Class F Insulation

| Hp | RPM | Amps | | | | Efficiency | | | | Torque | |
|---|---|---|---|---|---|---|---|---|---|---|---|
| | | 4/4 | 2/4 | 1/4 | OL | 4/4 | 3/4 | 2/4 | 1/4 | FL | Wk² |
| 150 | 3570 | 161 | 88.1 | 56.9 | 654 | 95.4 | 95.5 | 95.1 | 92.6 | 221 | 19.2 |
| | 1785 | 163 | 88.6 | 58.2 | 615 | 96.2 | 96.4 | 97.0 | 94.1 | 441 | 49.2 |
| | 1190 | 170 | 101 | 74.1 | 607 | 95.4 | 95.6 | 95.5 | 93.4 | 662 | 78.3 |
| | 887 | 179 | 104 | 75.3 | 614 | 93.7 | 94.5 | 94.8 | 93.3 | 887 | 97.8 |
| 200 | 3568 | 214 | 113 | 71.9 | 892 | 95.3 | 95.5 | 95.0 | 92.3 | 294 | 23.4 |
| | 1785 | 217 | 118 | 76.7 | 797 | 96.1 | 96.3 | 96.1 | 94.1 | 586 | 67.1 |
| | 1187 | 226 | 127 | 91.0 | 766 | 95.6 | 95.5 | 95.1 | 93.2 | 884 | 92.8 |
| | 888 | 251 | 156 | 122 | 800 | 94.8 | 95.2 | 95.1 | 92.8 | 1183 | 117 |
| 300 | 3571 | 318 | 170 | 104 | 1274 | 95.8 | 96.0 | 95.7 | 93.5 | 441 | 31.1 |
| | 1785 | 326 | 179 | 118 | 1224 | 96.4 | 96.7 | 96.6 | 94.8 | 882 | 86.9 |
| | 1186 | 338 | 185 | 123 | 1127 | 95.7 | 95.7 | 95.6 | 93.9 | 1327 | 110 |
| 400 | 3570 | 420 | 220 | 133 | 1678 | 96.2 | 96.4 | 95.9 | 92.9 | 588 | 39.0 |
| | 1784 | 431 | 224 | 135 | 1689 | 95.8 | 95.9 | 95.4 | 92.8 | 1178 | 89.1 |
| | 1185 | 449 | 228 | 138 | 1740 | 95.8 | 96.1 | 95.3 | 92.7 | 1774 | 118 |

The motor torque is

$$(160\,\text{Hp})(5255)/310\,\text{rpm} = 2710\,\text{lb-ft}$$

The drive's output is

$$0.14197\,10^{-3}\,(310\,\text{rpm})(2710\,\text{lb-ft})/(0.85\,\text{motor efficiency}) = 140.3\,\text{kW}$$

The power input to the drive is

$$(140.3\,\text{kW})/(0.94\,\text{drive efficiency}) = 149.3\,\text{kW}$$

The rectifier power factor is 95%.
The kVA input is 157 kVA.
As shown here, the motor's reactive current doesn't enter into the kilowatt load. (Motor reactive current has a small loss element in the drive efficiency). The drive's input power factor is used to calculate input to the drive.
A drive's output is composed of three torques:

1. An accelerating torque to bring the fan to operating speed (J torque)
2. A load torque at operating speed (for a fan, the air torque)
3. A dynamic torque that restores operation during a disturbance (dynamic torque)

The faster the starting, the stronger is the accelerating torque. It goes to zero when operating speed is reached. The load torque is the continuous torque at operating conditions. For a fan, it starts at zero, climbs as the square of speed when starting, and then loads the drive at operating value. The dynamic torque restores the fan to normal operation. A momentary flow blockage or other such event would disturb operation.

## Starting a Fan

Fans in industrial plants are often redundant. They are mounted side by side and ducted together. To prevent reverse flow when one fan is off, dampers are installed in each fan's outlet. A damper opens automatically by the developed pressure as soon as the fan is started. If one fan is running and the second fan is started, until the pressure developed by the starting fan exceeds the pressure of the running fan, the damper of the starting fan remains shut.

There is no flow pumped through the fan when it is blocked while starting. The air within the fan's housing is mixed by the rotor's impeller. Only differential pressure is about the same as developed by a starting unblocked

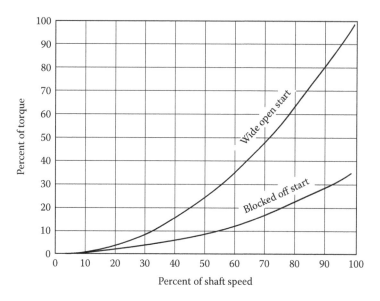

**FIGURE 2.2**
Torque–speed curve for a fan or air compressor.

fan at the same shaft speed. This blocked off start is shown on the plot of Figure 2.2. In this plot, the fan's torque is shown at beginning and operating speeds for an unblocked and blocked fan. (It is zero at zero rotation.) Once the fan is started, this torque is the total *continuous* operating air torque. This load torque is in addition to the torque supplied by the motor to accelerate the shaft.

The pressure to overcome the reverse-flow damper's restraining pressure is a required overpressure that opens the damper. Once overspeed rpm is reached, sufficient to open the damper, the flow is

a. Unblocked, and paralleled full flow commences, with the motor gaining parallel-load current.

b. The overspeed required to open the damper is then reduced to parallel-load rpm.

c. The overspeed torque is a dynamic torque.

d. Parallel-load flow is about half the flow if the fan were not operating in parallel, but operating by its self.

## Starting Ramp Time

In this discussion, the drive is configured to be in torque control mode. Ramp time is also configured in the drive's firmware. For fan duty, constant acceleration is utilized for the ramp-up in speed. The ramp time controls both acceleration and deceleration when reducing speed. Current is regenerative during coast down. Just as during acceleration, the inverter shapes regenerative current. The regenerative current is the same as when accelerating, only shifted 180°. Deceleration has the same ramp time as acceleration.

During starting, a constant torque applies a constant acceleration. This is the J torque. The drive applies to the shaft J torque plus the air torque. Unlike a conventional across-the-line starter, where a constant voltage is applied to the motor, a drive increases applied motor frequency in the current to the motor from zero to full rpm. *There is no locked rotor current.* The square of current is roughly proportional to applied torque, starting with the J torque value. The air torque climbs to the full-load value, typically as shown in Figure 2.2. The total torque demanded of the motor is the sum of the accelerating torque and the air torque. Maximum torque occurs when the drive is accelerating near full-load rpm. Accelerating torque is applied according to the configured ramp time, $\mathbf{T} - t_0$. Faster ramp time demands more J torque. This is also the J torque value used by the drive to effect any change in speed. Figure 2.3 illustrates starting torques.

During starting, if a disturbance happens, the disturbance's dynamic torque is also added to the total torque shown in Figure 2.3. If the disturbance happens toward the end of the start, additional maximum torque is required. The total torque, with the added dynamic torque, will depend on the dynamic torque's amplitude. Usually, its amplitude is a multiple of the J torque.

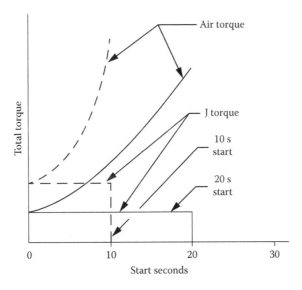

**FIGURE 2.3**
Starting torques of a motor driving a fan.

## Mechanical Resonance

Mechanical resonances happen during rotation at some fan shaft speeds. The fan manufacturer measures these resonances during testing. Drives do not operate at these speeds. The drives are configured with the measured resonant frequencies and do not operate at these speeds.

Who hasn't seen a top-loading washing machine starting to spin-cycle but stuck at a thumping-away, slow speed? The clothes are all on one side in the tub, and the machine either trips or continues to loudly thump away. With the machine turned off and its lid raised, the wet soggy clothes can be evenly rearranged around the tub's edge. This eliminates the imbalance and thereby eliminates the unwanted loud mechanical resonance of this imbalanced load. Now the limited torque of the motor isn't called upon to supply the greatly added power demanded by the slowly spinning imbalanced soggy clothes. The motor smoothly spins the clothes up to its high speed and completes the spin cycle.

Added torque is applied to quickly accelerate through a resonant shaft speed. At a resonant shaft speed, a flow of energy into the vibrating mechanical parts demands still more torque because this energy flow is supplied by the shaft.

The band of shaft speeds characteristic of a mechanical resonance depends on the vibrating structure. The drive must pass through the resonant shaft speed and proceed with further accelerating the shaft. Hence, with the increased flow of energy into the shaft to feed the resonance and the need to quickly accelerate the shaft beyond the resonance, a dynamic torque is demanded. The air torque at the resonant shaft speed is the same; whether resonance is happening or not, the drive responds to a mechanical resonance by adding a dynamic term to the steady value of shaft acceleration.

The manufacturer has stiffened and balanced rotating members to reduce resonance and shift resonant speeds above operating speed. However, some resonances are inherent. Generally, the flow of added energy to the resonating members, and the need to quickly accelerate through the band of shaft speeds, requires some multiple of the value of J torque.

In measuring the resonance during factory testing, the manufacturer may have measured the motor current before and while the resonance occurs. If so, the square of the current increase is roughly proportional to the torque required to feed the resonance. (This assumes that the resonance doesn't slow shaft speed.) If the manufacturer cannot supply this measured resonance-feeding change in motor current, a value of triple the J torque will generally cover dynamic torque due to resonance.

The drive's default value for maximum torque configuring is a boundary for the sum of the total torque's parts. *Remember that the size of the UPS is set by the configured torque limit setting. See the following example for sizing the UPS.* Also the reverse is true: the torque limit set in the drive is determined by the size of a UPS source.

The assumption that the drive's power source is almost unlimited, compared with the rating of the drive, is found throughout manufacturers' literature. But the drive's power source is highly limited if the drive's supply is a UPS. Instructions in a drive manufacturer's literature must be carefully reviewed, and drive default values and configuring, all must be carefully set. A UPS has two operating domains. If load is 150% or less, the UPS is narrowly voltage regulated, about 1%. Thus, a small change in load current with a minute change in voltage looks to the drive like an almost infinite source.

The other operating domain is excessive overload and short circuit. If short-circuited, the UPS can supply 10 times rated output current. However, this current is supplied only for a rated three *cycles* and is highly distorted with harmonics.

Factory tests of a 200 kVA 480 V three-phase UPS showed the quality of the voltage and current plot as virtually harmonic-free. The test results show a total harmonic value of 4.51%. The short-circuit voltage and current, however, have maximum distortion. The UPS can withstand overloads of 150% for more than 30 s, longer than a drive. It is rated for 1 min with a 125% overload.

### Response Time of a UPS with a Drive's Load

Electronic delay in signals results from the filtering in a drive's rectifier and in the DC bus filtering. However, this delay is negligible compared to the few seconds of a dynamic torque demanded by the load. The motor's response added to a drive's response is still too short a delay to affect a mechanical load. Added to the drive's and motor's response is that of the UPS output and input. But again, this sum is negligible compared with the dynamic torque's demand. From a dynamic torque that lasts for several seconds, a UPS and drive see this load as a steady-state demand.

To meet an increase in shaft load demand, the drive's inverter increases current to increase flux in the motor's air gap. This cascades back to the UPS's input rectifier. Disturbances demanding a dynamic torque response are subject to the response time of the fan's inertia. This inertial response is measured in seconds. Because the cascaded response time of the electronics is less than a cycle, the fan sees this increase in motor torque as a practically instantaneous change.

### Fan Start Time and Torque

As stated before, there are three torques involved: the J torque, the air torque, and the dynamic torque. The faster the required start time, the more required accelerating J torque. Because the drive is supplied by a UPS, the current demanded by the drive must be limited. This limit determines the size of the UPS. This limit is set as a configured parameter in the drive's firmware. Usually, the drive's capability (with margin) is in the default setting of the torque output; 300% of rated is a common default limit. The sum of the three torques must not exceed the configured parameter of torque limit.

Conversely, the configured torque limit must be set high enough to accommodate the three torques. Maximum torque is roughly proportional to the square of the maximum current that the UPS can supply. Total motor torque within a drive is calculated from

$$\tau = \frac{3}{2}\frac{P}{2}\frac{L_m}{L_r L_{s'}} \text{ß}_r \text{ß}_s \sin\Phi$$

where
P is the number of poles
$L_m$ is the mutual inductance
$L_r$ and $L_{s'}$ are the rotor and stator leakage inductances
$ß_r$ and $ß_s$ are the rotor and stator fluxes
$\Phi$ is the angle between them

Stator current and rotor back emf are adjusted to provide the required torque. (The drive is operating in configured direct torque control.)

As with torque, the slip rpm is also calculated within the drive. Slip is calculated from

$$\omega_{slip} = \frac{2}{P}\left(\frac{3}{2}R_r\frac{\tau}{\psi_r^2}\right)$$

where
$\omega_{slip}$ is in rad/s
$R_r$ is the rotor resistance
$\tau$ is the torque
$\psi_r$ is the rotor flux, as $ß_r$ earlier

Total torque during starting is the sum of the accelerating torque (J torque), the increasing torque of the fan's air load (air torque), and any dynamic load that may happen during the start interval. If a given fan is to be started in 5 s, a large J torque is required. This torque value is twice the value of a J torque for a 10 s start and four times the value for a 20 s start. With dynamic torque a multiple of J torque, the sum of torques for a 5 s start is dominated by the J torque and dynamic torque. *They can require a larger drive and UPS.* Start time must be carefully adjusted to the needs of the system.

Start times of about 60 s are usually acceptable. The fan was off before a start is required. If the need for flow can be anticipated, a longer start time can be configured. If for redundant fans, airflow is critical and failure of the operating fan sequences starting the standby fan, falling shaft load can be sensed in the failed fan's drive, and can signal start of the redundant fan.

During the time of reducing flow to a minimum tolerable value and using the failed fan's slowing time, a start time beyond 5 s can be utilized.

The air torque load depends on the airflow. The Fan Laws state that airflow is proportional to the shaft speed and changes in the airflow system (changes in valves and dampers). If the airflow system doesn't change (no change in dampers and valves), the Fan Laws also state that airflow torque is proportional to the square of the rpm.

## Motor Efficiency Requirements

Motor efficiencies at 60 Hz operation are high, usually above 90% for loads from one-fourth to full load (not true for smaller motors) (see Table 2.1). Motor efficiency at light load, where a variable-frequency drive may be required to operate, has a primary influence on sizing the UPS and especially on sizing its battery. A motor operating at light load at 60 Hz has less efficiency than at half and full load because the motor's losses are proportionately greater than losses near full loads.

Manufacturers' data typically list efficiencies from one-fourth to full load at 60 Hz. What if the current's frequency is less than 60 Hz? With the drive supplying current at 20 Hz, what does the motor performance look like? Table 2.2 shows the estimated performance of a 150 Hp 3600 rpm motor at 20 Hz. The second and third columns are based on the approximations that over the load range

a. The motor's torque at 20 Hz is the same as at 60 Hz.

b. The slip rpm at full load is the same as at 60 Hz at 150 Hp.

c. The slip rpm is proportional to the load.

These approximations are usually good for the range of frequencies from 30% to full load. If high accuracy is needed, the manufacturer will have to be consulted. He may be required to perform factory tests.

**TABLE 2.2**

Rough Motor Characteristics at 20 Hz

| 150 Hp 460 V TEFC 3600 rpm motor | | | |
|---|---|---|---|
| See values in Table 2.1. | | | |
| Synchronous speed: 1200 rpm | | | |
| Percent Load | Hp | RPM | Percent Efficiency |
| 100 | 50 | 1170 | 95.4 |
| 75 | 37.5 | 1177.5 | 95.5 |
| 50 | 25 | 1185 | 95.1 |
| 25 | 12.5 | 1187.5 | 92.6 |
| 11 | 5.5 | 1197 | ? |
| 0 | 0 | 1200 | 0 |

In the fourth column, efficiencies are copied from those listed in Table 2.1 for the 150 Hp 3600 rpm motor. These values too are approximations, when used for the values at 20 Hz. In this reduced-frequency table, the efficiency is almost constant, but as load decreases, the efficiency is beginning to fall off at one-fourth load. Here the efficiency is 92.6%, reducing from 95.3% at higher loads. At no load, the efficiency is zero. With a fan loading the motor, the full-load airflow at 1170 rpm is now 33% of the flow at 3570 rpm because flow is proportional to shaft speed. For this fan, at this rpm and flow, its Hp drops off as the cube of the shaft speed. Then if the fan loads the motor to 150 Hp at 3570 rpm, at 1170 rpm the fan Hp would be only 3.5%, or 5.28 Hp. With this load, the motor's slip rpm decreases, increasing rpm slightly, and the load adjusts to about 5.5 Hp at about 1197 rpm. This load is 11% of 50 Hp, as shown in Table 2.2.

If a linear interpolation is used to find the efficiency at 5.5 Hp, the value would be 40%. But efficiency would be higher if a curving plot of efficiency versus load were used to interpolate the value. In this plot, in the region from zero load to 12.5 Hp, a curved plot of efficiency from zero to 92.6% would be depicted. Interpolation on this plot at 5.5 Hp would yield a value of about 70%.

This is a large difference in the two interpolated values. In finding the size of the UPS's battery, as shown in the following example, these two values dictate two very different battery physical sizes. In the example, a motor efficiency at light load and slow rpm is 52%. If an accurate battery size is essential, the motor manufacturer should be consulted for the efficiency.

## An Example

The rating and size of a UPS is to be found. The UPS is to supply a variable-frequency drive that is supplying a motor driving a fan. Normally, the drive is supplied from a plant 480 V source. But with loss of normal power, the drive is switched to the UPS. For this operating mode, the required flow is 40% of normal flow, or 640 cfm.

The fan

1. At 1787 rpm, the fan supplies 1600 ft³/min to its process system.
2. The load horsepower at this flow is 305 Hp, which is the worst-case fan load.
3. The fan's required starting time is 10 s.
4. The fan's Wk² is 1660 lb-ft².
5. The fan has a mechanical resonance at 1620 rpm.

6. At 1620 rpm, a sustained torque of 1060 lb-ft was measured with the fan loaded.
7. The fan's load at 640 cfm is 19.5 Hp at 715 rpm.
8. This flow must be maintained for 20 min.

## The motor

1. The motor is rated 350 Hp at 1785 rpm and 460 V.
2. The motor's rated torque at 1785 rpm is 1029 lb-ft.
3. The motor's Wk$^2$ is 87 lb-ft$^2$.
4. The motor's full-load efficiency at 460 V is 96.2%.
5. The manufacturer states that at 715 rpm and 19.5 Hp the motor efficiency is 52%.

## The drive

1. The drive is rated 400 Hp at 480 V and 1800 rpm.
2. The drive's torque at 400 Hp and 1800 rpm is 1168 lb-ft.
3. The drive has an overload rating of 150% for 20 s.
4. The input power factor is 95%.
5. At light load, the drive's efficiency is 92%.
6. At full load, the drive's efficiency is 94%.

## The UPS

1. When operating from the UPS, the fan supplies 640 cfm to its system at 19.5 Hp.
2. The inverter efficiency is 93%.
3. The overall efficiency is 91% at a power factor of 94% at full load.
4. The overload capacity is 150% for 60 s.
5. The battery is a lead–acid valve-regulated type discharged to 1.75 V/cell.
6. The battery will require 190 cells (380 VDC) to supply the UPS's inverter that supplies the drive at its rated 480 V.

## Normal Operation

At 305 Hp, the motor slip is less than 15 rpm at 350 Hp. The slip is proportional and is

$$(305/350)(15\,\text{rpm}) = 13\,\text{rpm}$$

Thus, the full-load rpm is 1800 − 13 = 1787 rpm, and this speed during acceleration is reached in the 10 s of start time. The acceleration is 1787/10 = *178.7* rpm/s, or 2.978 *rps*/s, or 2.978 rev/s². Thus

$$\alpha = 178.7 \text{ rpm/s}$$

The inertia is the sum of the fan's and the motor's Wk², divided by g:

$$\left(1660 + 87 \text{ lb-ft}^2\right)/32.16 \text{ ft/s}^2 = 54.32 \text{ lb-ft-s}^2/\text{rev}$$

J = 54.32 lb-ft-s²/rev. And with $\tau = J \alpha$

$$\tau = (2.978)(54.32) = 161.8 \text{ lb-ft for the 10 s of acceleration.}$$

J torque is 161.8 lb-ft for 10 s.
   At full load, 350 Hp, the motor's torque is 1029 lb-ft at 1785 rpm.
   At 305 Hp and 1787 rpm

$$\tau = \text{Hp}(33,000)/2\Pi \text{ rpm} = (305)(33,000)/(6.28)(1,787) = 896.9 \text{ lb-ft}$$

The fan's resonance was measured at 1620 rpm, and the measured torque was 1060 lb-ft. This is the sum of the air torque and the resonant torque at the tested factory conditions. At this rpm, the air torque cannot be known because, for the fan in the factory, although loaded, only the sum of the air torque and dynamic torque was given by the manufacturer. The air torque during the test can be guessed at, however. The guess is that the fan was loaded as in the operating conditions herein.

   At 1620 rpm, the air torque would be (1620/1787)² (896.9 lb-ft) = 737 lb-ft, if operating the fan at the same loading as herein. Guessing, then, that the manufacturer tested the fan with a load of 700 lb-ft of air load when operating just below, and just above, 1620 rpm. The manufacturer measured 1060 lb-ft of torque at this resonant speed in the factory. Then the estimated dynamic torque during the resonance was 1060 − 700 = 360 lb-ft. Adding a small margin, the dynamic torque when operating at resonant rpm is estimated at 400 lb-ft.

   Herein, this torque value is further used for *any* dynamic torque that may happen during operation. And the worst-case operation would be if a disturbance causing the dynamic torque happens just as full rpm is reached during starting.

   Summary

| | |
|---|---|
| Continuous shaft torque at 1787 rpm 305 Hp | 896.9 lb-ft (air torque) |
| Starting torque for 10 s start | 161.8 lb-ft (J torque) |
| Worst-case estimated dynamic torque | <u>400.0 lb-ft</u> (dynamic torque) |
| Total peak torque | 1458.7 lb-ft |

The torque limits of the supply equipment are

| | |
|---|---|
| Motor's rated continuous torque | 1029 lb-ft |
|    Breakdown torque at 200% | 2058 lb-ft (2058 > 1459) |
| Drive rating at 400 Hp and 1800 rpm | 1168 lb-ft |
|    150% for 20 s | 1752 lb-ft (1752 > 1459) |
| Dynamic overload of drive (1459/1168)(100) = | 125% for 10 s (125% for 10 s < 150% for 20 s) |

Hence, the capacity limits of the supply equipment are not exceeded.

### Time Where Load Exceeds Steady Torque

The steady torque is 896.9 lb-ft. The time when the accelerating shaft reaches this torque is calculated from when the accelerating shaft reaches the speed where the torque is this value. When the increasing air torque and the J torque reach 896.9 lb-ft, the air torque has a value of 896.9 − 161.8 = 735.1 lb-ft. Torque is proportional to the square of speed: speed = $(735.1/896.9)^2$ (1787 rpm) = 1200 rpm. Accelerating at 2.978 rev/$s^2$, or 178.7 rpm/s, the time to reach 1200 rpm is 1200/178.7 = 6.7 s. And with 10 s of acceleration, the overload lasts for 10 − 6.7 = 3.3 s.

The drive is capable of 150% for 20 s. This is a capability of 3000%-s. The overload is 127% for 3.3 s, or 419%-s, which is well within the drive's capability.

### Setting the Drive's Torque Limit in the Drive's Firmware

The torque limit is set above any operating mode that the drive may encounter. It must be higher than the 1459 lb-ft, stated in the summary of "Normal Operation" section. The dynamic torque is assumed to be 400 lb-ft, but what if a greater torque is encountered? If an allowance of 800 lb-ft is used, then a torque limit must be higher than 1459 lb-ft, plus another 400 lb-ft, or 1859 lb-ft.

A common default setting for torque limit in variable-speed drives is 300% of rating. The said drive has an Hp rating of 400 Hp at 1800 rpm. Its rated torque is 1168 lb-ft. The minimum dynamic limit of 1859 lb-ft is 159% of the drive's full-load rating, well below 300%. Adding 41 lb-ft to the 1859 lb-ft, for a 2% margin, yields a 1900 lb-ft setting for the drive's configured torque limit setting.

As seen in the following, this 1900 lb-ft limit sets the size of the UPS.

### Sizing the UPS for Its Load

The fan's load at 715 rpm and 640 CFM is 19.5 Hp. The manufacturer gives the motor efficiency at this load as 52%. The motor's input is 19.5 Hp/52% Eff, or 39 Hp. With this load, the drive is at low load and has an efficiency of 92%.

| | |
|---|---|
| 39 Hp/0.92 = 42.4 Hp, and times 0.746 = 31.6 kW supplied to the drive. | |
| The dynamic momentary torque is | 400.0 lb-ft |
| If a need for starting, the J torque is | 161.8 lb-ft |
| Total momentary torque | 561.8 lb-ft |
| The continuous air torque is | |
| 19.5 Hp (33,000)/2Π 715 rpm = | 143.3 lb-ft. |
| The total torque is | 705.1 lb-ft. |

The total load torque at 715 rpm is 705.1 lb-ft.

If this motor's torque stays constant as the supplied current's frequency is reduced, the torque would stay at 1029 lb-ft. Then the total torque of 705.1 is about 80%. Like torque, if the efficiency also stays constant as the supplied current's frequency is reduced, the efficiency at 80% load is about 96.3% in this motor. For the total torque, the efficiency is about 96.3%. But for the continuous torque, the load is only about 14%. Here, the manufacturer states the motor efficiency is 52%. The motor's momentary Hp is

$$2\Pi(561.8\,\text{lb-ft})(715\,\text{rpm})/33,000 = 76.4\,\text{Hp}$$

The motor's momentary input is

76.4 Hp(0.746)/0.963 motor efficiency

= 59.4 kW motor's momentary input and the drive's momentary load

The motor's continuous input is

19.5 Hp(0.746)/0.52

= 29.1 kW motor's continuous input and the drive's continuous load

These are light loads on the drive; hence, its efficiency is 92%. Dividing this efficiency into the drive's loads determines the UPS loads:

| UPS load at 715 rpm power input | |
|---|---|
| Momentary | 64.6 kW |
| Continuous | 31.6 kW |
| Total | 96.2 kW |

The total UPS load is 150% of its rating. Were the rating 71.3 kW, its momentary capacity would be 106.9 kW, and 90% would be 96.2 kW. Then the continuous load is 31.6 kW/71.3 kW, or 44%. Thus, the UPS's minimum size is 71.3 kW.

Adding a slight margin to the momentary and continuous loads, the operating load becomes

| Continuous | 35 kW |
|---|---|
| Momentary | 70 kW |
| Total | 105 kW |

With the maximum torque configured in the drive at 1900 lb-ft, and with a load rpm of 715 rpm, also configured in the drive, the load is

$$(6.28)(1,900)(715)(0.746)/33,000 = 192.9 \text{ kW}$$

but this would be for a momentary load.

$$192.9 \text{ kW}/150\% = 128.6 \text{ kW continuous raring}$$

The specified capacity of the UPS will be

| Continuous | 130 kW |
|---|---|
| Momentary | 195 kW for 60 s |

With the drive's input power factor of 94%, $128.6/0.94 = 136.8$ kVA. Thus, with margin, the drive rating is 140 kVA.

### Sizing the UPS Battery

If the momentary load of 70 kW mentioned earlier were to last a full 10 s, the kilowatt-minute load would be 11.7 kW-min. The continuous load, 35 kW, lasts for 20 min, which is a load of 700 kW-min. The momentary load is (11.7/700) 100, or 1.7%. This is negligible.

The battery is sized to supply only the fan at 640 cfm and 715 rpm for 20 min. This is with efficiencies of

| Motor | 52% |
|---|---|
| Drive | 92% |
| UPS inverter | 93% |

The battery supplies the cascaded load as follows:

| Fan load | 14.6 kW (19.5 Hp) |
|---|---|
| Motor load at 52% | 28.1 kW |
| Drive load at 92% | 30.5 kW |
| UPS inverter at 93% | 32.8 kW |

**TABLE 2.3**

Battery Characteristics Valve-Regulated Lead–Acid Cell 1.75 V/Cell kW/Cell at 25°C

| Module | Minutes | | | | | |
|--------|------|------|------|------|------|------|
|        | 1 | 5 | 10 | 15 | 20 | 25 |
| MF250  | 0.367 | 0.271 | 0.201 | 0.152 | 0.138 | 0.110 |
| MF275  | 0.533 | 0.423 | 0.307 | 0.244 | 0.174 | 0.154 |
| MF2100 | 0.710 | 0.564 | 0.409 | 0.325 | 0.268 | 0.232 |
| MF2125 | 0.808 | 0.666 | 0.501 | 0.406 | 0.339 | 0.290 |
| MF2150 | 0.969 | 0.799 | 0.601 | 0.487 | 0.407 | 0.309 |

The battery must supply a 32.8 kW input into the UPS inverter for 20 min.

In the beginning of this example, it states that the battery consists of 190 cells discharged to 1.75 V/cell. These cells are connected in series. Thus, each cell in the string must be capable of delivering 32.8 kW/190 cells, or 0.173 kW/cell, for 20 min. The required cells must each include more than 0.173 kW for 20 min. The load must be adjusted with a capacity factor and a temperature factor that increase the cell's needed capacity.

The capacity factor adjusts for the cell's decrease in capacity with time. A multiplier of 1.2 is typical. The temperature factor adjusts for the normal operating temperature of the room that the battery is located in. A room temperature of 25°C requires no adjustment. If a warmer room temperature applies, such as 30°C, then a factor of 1.15 is appropriate. An average temperature for a warm day should be applied. A 30°C average would include a 40°C afternoon temperature. Such a day is rare. If the room is air conditioned, no temperature adjustment is recommended. Using adjustments of 1.2 and 1.1, or 1.32, and multiplying by 0.173 kW/cell, the battery's cell capacity should be 0.228 kW/cell. Table 2.3 lists cell capacities that are typical for battery manufacturers' products.

Looking down the next-to-last column for 20 min capacity cells, a module of 0.268 kW/cell is shown. This is an MF2100 module in the table. As the table's heading states, these cells are valve-regulated lead–acid type discharged to 1.75 V/cell at 25°C.

This MF2100 module contains three cells. When mounted, the 64 modules containing 192 cells are stacked to about 8 ft high and are about 5 ft wide and 3 ft deep. Usually, the battery is mounted next to or near the UPS.

## Light Loads on Large Motors

As discussed under *Motor Efficiency* before, a motor's efficiency has a direct effect on sizing the battery. In the fan example mentioned before, a motor efficiency of 52% was given by the motor manufacturer. If a higher efficiency were known, look at what happens.

With an efficiency of 85%, the required battery load becomes

| | |
|---|---|
| Fan load | 14.6 kW (19.5 Hp) |
| Motor load at 85% | 17.2 kW |
| Drive load at 92% | 18.7 kW |
| UPS inverter at 93% | 20.1 kW |

The battery must now supply a 20.1 kW input to the UPS inverter for 20 min. And with the same adjustment to cell capacity of 1.32 used before, the required cell capacity becomes

$$(20.1/190)(1.32) = 0.14 \text{ kW/cell}$$

In Table 2.3, the next smaller module, MF275 with 0.174 kW/cell, is now required. This module contains four cells, and this battery has 48 of these modules containing 192 cells. Its dimensions are 8 ft high by 36 in. wide by 24 in. deep (and costs less).

Examining in Table 2.1 the four columns under *Efficiency*, for the four sizes of large motors, there is a remarkable sameness to the values in these columns. Each of the four motors has three and four windings, yielding the motors' speeds as shown. Yet in all these windings, stators, and frames, the efficiencies are about the same. From loads of full to half, the efficiencies are all above 95%. At one-quarter load, the efficiencies begin to decrease, and all values are fairly uniformly dropping between 1% and 2%.

Figure 2.4 shows this characteristic of these motors. Manufacturers rarely address loads less than 25% because use of smaller motors for such loads is

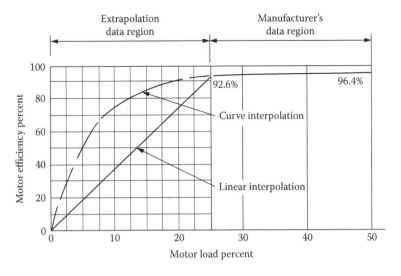

**FIGURE 2.4**
Large motor efficiencies at loads below one-fourth.

appropriate. But drives can operate at synchronous speeds down to 1200 rpm for 3600 rpm motors. Here, motor efficiencies at light loads are needed. The fan example stated before illustrates this need. *And efficiencies remain about the same with reduced synchronous speeds.* Therefore, Table 2.2 and Figure 2.4 are appropriate for reduced synchronous speeds.

In Figure 2.4, comparing the manufacturer's supplied efficiency value of 52% in the example mentioned earlier, with the linear interpolation value, it is obvious how the manufacturer obtained his 52% value. The curve extrapolates the efficiencies discussed in Table 2.1 in the region shown. The two interpolations are at a value of 14% motor load. This is 19.5 Hp, with 720 rpm synchronous speed, and 140 Hp at full-load, for the 350 Hp motor.

# 3

## Storage Batteries

Intense battery development is currently under way to find a cheap, light-weight, high-energy battery for cars, airplanes, and solar and wind power plants. Hopefully, the developed battery will also be capable of many, many charge–discharge cycles. In early flights while introducing the Boeing 787 Aircraft, batteries on board during flight ignited and caused fires and alarms, as well as caused cautions over the safety of this new airplane. A redesign fixed the problem. Speculated at the time was that the engineering was faulty. The battery overheated and ignited because in the installation's design, heat dissipation was not properly addressed, or reduced convection at altitude wasn't accounted for (convection reduces above 1 km). Battery heat dissipation and convection are addressed in this chapter.

The most numerous type of battery today is the lead–acid battery. It is in every industrial plant, as well as every car and truck in the world. In developing the Apollo Lunar Module, six batteries formed its power sources, four in the descent stage, which remained on the moon, and two in the ascent stage that returned the two astronauts to the command module orbiting overhead. These batteries were silver-oxide type, the most energetic available at the time. Highly expensive, they could be charged and discharged only once. But once was all that the mission required. They weighed less than the fuel-cell systems used in the command module and were chosen because of their lighter weight.

The lead–acid battery is discussed herein because of its most widespread use. Despite its low cost, it is still too costly for solar and wind power storage. The cost of storing megawatt-hours has to compete with fossil power generation. Despite subsidy, the cost of generating and storing renewable power can't compete with a base-load fossil plant yet. The lead–acid battery is discussed as follows because it is the most common, and also because it illustrates the principles of using any kind of battery.

## The DC System

Switchgear, transfer switches, and the like need to be controlled when everything goes bad. In the example of Chapter 2, a transfer switch has to work during a power outage. It needs to operate when the building power is dead. Medium-voltage switchgear must operate when there are faults on a plant's heaviest feeders. The entire plant system's voltages are very low during the time the protective relaying senses the fault, and the switch-gear's breaker trips to isolate the faulted feeder from the rest of the supply system. In other words, when things are bad (when Murphy is at work), the controls must operate.

Large industrial plants first were designed in the early 1930s. The classic source for electrical control power was the DC system. It consisted of a panelboard filled with 2-pole circuit breakers feeding the ungrounded control circuits. There were three sources of DC power, a stationary bat-tery and two redundant battery chargers, each fed from two different plant sources. All were grouped in the electric room, with the battery in an adja-cent room. Because the battery's terminals were exposed, the battery had to be isolated.

Normally, the chargers powered the system. One would be set with volt-age slightly higher than the other so that load sharing wasn't a feature. The voltage was set to the battery's float value, about 2.2 V/cell. The bat-tery was usually wet-cell lead–acid with 60 cells. The nominal system volt-age was 125 VDC, Edison's old voltage. This took advantage of all the solenoids, contactors, etc., on the market and is a standard value today. In steam power plants, the largest load was the turbine generator shaft's emergency lube pump.

Today, little has changed in this system. The stationary wet-cell battery has proved to be still the most reliable. The modern valve-regulated type doesn't have the cell life of the wet cell. Also, changing out a cell is difficult in the stacked arrangement. But the chargers are much better with solid-state rectification. Also, the monitoring and alarms are now centralized, which is a main improvement. Plants have burned down because the monitoring and alarm panel wasn't located where the operators were.

Lately, designs have been substituting UPSs for the DC system. This is not desirable for three reasons:

1. The UPS battery is of the valve-regulated type and is not as reliable as the stationary wet-cell battery.
2. There is no charging and supply redundancy.
3. Some controls are so fast acting that trips happen during switching between UPS sources.

One small generating plant fueled with natural gas had blocking valves in its fuel system. The valves were electric and were connected to a UPS. Cycling the UPS output between the inverter and the UPS's backup tripped the plant, because the valve solenoids acted in less than a quarter cycle. The valves were changed to 125 VDC and were connected to the plant's DC system.

The DC system is an essential emergency system. Its continuous load is very small, some indicating lights with a few monitors and relays. It is there for trouble and is sized for trouble. Standards dictate how the system is sized. Several myths have grown up about how the battery room is ventilated and wired, and how the panelboard's circuit breakers are sized.

The battery room scares designers. The battery gives off hydrogen gas. No one seems to calculate how much. It is a very small amount. They overventilate and they classify the wiring, all because they don't know how to do otherwise. They provide unnecessary shower apparatus when only a water bucket is needed. Even though a broken cell has too little electrolyte in it to reach the drain, they provide a chemical-resistant floor drain. They fail to anchor the rack to the floor. Too often the panelboard is oversized. Ignored is the fault current that the battery can deliver, often less than 5 kA. (Look in a catalog at DC breakers and their interrupting ratings: you will find 5 and 10 kA.) In a medium-voltage circuit breaker, the close circuit is protected by a 5 A fuse. A coordinating 10 A fuse feeds the paralleled close and trip circuits. The breaker has its closing circuit's spring charging momentary load, and a few small indicators and relays are the continuous loads.

*Overload protection of conductors is not required!* Only fault-current protection is required, like fire pump wiring. This means that a 12 AWG wire size, feeding the switchgear, can have a 30 A DC breaker supplying it (the breaker can interrupt the battery fault current, and it coordinates with the 10 A fuse in the switchgear circuit breaker). To coordinate with this breaker size, a 100 A breaker in the panelboard for connecting the battery is all that is needed. The panelboard bus size could be 60 A because the continuous load is so small. The feeder from the battery must be sized for voltage drop when the control circuits operate. Its main load is the battery's charge current, but this is not a continuous load. Its 2 h average value is modest, drawing a short-time heavy load in the beginning. The charging current would not open a 100 A breaker. All DC circuits are sized for voltage drop. Because the DC wiring is emergency wiring, it must be separated from the building wiring. This is often unrecognized.

The circuit breakers for the chargers must pass the charge current. These could be 50 A breakers. The chargers cannot supply fault current and need not coordinate with the 30 A branch breakers.

Calculating rms values of charge current profiles, and coordinating with the circuit breaker's DC trip curves, is appropriate for sizing charger circuit breakers.

It is about four times harder to interrupt a DC current than an AC current, because the AC current passes through zero twice a cycle. A 22 kA interrupting capability for AC might be a 5 kA DC capability. However, standards dictate whether or not a manufacturer's AC rated breaker can be derated and used in a DC circuit without the breaker being listed, or without the manufacturer's permission.

## Valve-Regulated Lead–Acid Battery

The lead–acid wet-cell battery consists of cells connected in series. In a wet-cell battery, each cell is made of a stack of lead-coated plates suspended in a sulfuric acid solution within a glass jar. The solution is the electrolyte. As well as the cell's terminals, included in the lid is a vent for generated hydrogen. Each plate has a calcium substructure that is porous with a maximum surface area for the deposited lead. On charging, the lead plates out on this surface area, and on discharging, the lead corrodes away into the sulfuric acid solution. The substrate's shape renders a maximum number of charge–discharge cycles, as well as giving the cell a rugged property. In the stack, every other plate is bonded together and bonded to one of the cell's two terminals in the lid. After assembly, a DC current is passed through the cell, which forms a positive-negative-positive function to the plate stack. To cover both sides of the positive plates, the outermost plates are negative, yielding one more negative plate than positive plates. Cells are connected in series to obtain the battery's voltage.

With practically all the lead ions in the electrolyte plated onto the positive plates, the battery is fully charged. Continuing to drive a charging current through the battery causes hydrolysis of the water in the electrolyte. Hydrogen is liberated on one plate and oxygen is liberated on the other plate at a rate depending on how much current is overcharging. These gases vent to the room.

If this current is not switched off, continued venting and an explosion of the hydrogen can happen. To quickly charge a battery, a high current is passed through during early charging and is reduced as the cell is recharged. Chargers have automatic control of the charge cycle, but there is a manual control of overcharge current for maintenance procedures. This control is a hazard. Continuous overcharge currents constitute a failure. Battery explosions are associated with abandoned premises and misused chargers.

In stationary service, a battery sits fully charged. It also is self-discharging in this state. A small equalization current (called a trickle charge) is supplied that offsets this discharge. Also, hydrogen is evolved at a small rate. If one were to list the chemical equations that describe a battery, there would be many. In the valve-regulated battery, the hydrogen and oxygen recombine within the cell.

A valve-regulated battery cell is dry inside. Instead of a bath of electrolyte, layered fiber glass cloths soaked in the lead–acid solution are stacked between each of the plates and are the electrolyte. The cell's construction contains the small amount of generated oxygen and hydrogen, and the two gases recombine within the cell. Over time, a residual amount of gases builds up the internal pressure. A vent valve releases this small amount at about 0.7 psi and reseals at 0.5 psi. About a cubic inch is vented typically about once every other month.

The cells are assembled in modules of three or four cells. The modules usually are also conveniently installed in stacks. The stack structures provide spacing between modules for convection currents, and to aid accessibility. Module faces having connection terminals are aligned, and the stack face with the connections is the front of the battery. To prevent contact, a clear plastic barrier covering the terminals is installed. Cells need replacing during the battery's lifetime, and this construction allows replacement. Wet-cell batteries have exposed live terminals and need isolation. Valve-regulated batteries do not.

## Battery Selection

Battery requirements define the selection. A battery's requirements are the ampere load to be supplied over a discharge period, and within limits of voltage and voltage regulation. Also, other requirements are the number of charge–discharge cycles over the life, weight and size, and cost. Comparison of the different types also plays in the selection.

A manufacturer may have three or four plate designs that differ in size. The basic performance of a plate is, if used in a battery, the plate can deliver a value of ampere-hours in an 8-h period. The positive plate has the basic capacity, with the negative plates on each side (all plates look the same) serving the discharge. The larger the area in the plate, the higher the amp-hours capacity. By stacking more plates in a cell, the cell can deliver more amp-hours from the paralleled positive plates in the cell. But there are practical limits to the number of plates. So for large batteries large plates are stacked. And for smaller batteries smaller plates are stacked in a cell.

Any device used for storing energy, be it a flywheel, or molten salt, or hot rocks, or compressed air, etc., is never fully discharged. Inherently, energy is withdrawn within usable limits, such as a slowing of spinning, or a decrease in temperature or pressure, etc. With a battery, its voltage decreases as it is discharged. The fully charged voltage is the upper value of usable range, and the depth of discharge, with its lower voltage, is the lower value. Different battery applications require different voltage ranges.

Thus, the manufacturer has to list his cells' performances by depth of discharge: for each discharge voltage, cell performance is listed by discharge time. Table 3.1 illustrates such a listing.

In Table 3.1, two batteries are listed, which use the same plate but differ in their number. The plate is a 50 A-h plate. In 1 battery, there are 10 positive plates, for a total of 21 plates. In the second battery, there are 12 positive plates, for a total of 25 plates. The listing is for a discharge voltage of 1.8 V/cell (i.e., all plates are in parallel in each cell).

**TABLE 3.1**

Battery Performance (VRLA Type at 25°C 1.8 V/Cell Discharge)

| Cell | Positive Plates | Amps at 120 h | Amps at 40 h | Amps at 8 h | Amps at 4 h |
|------|-----------------|---------------|--------------|-------------|-------------|
| M50S21 | 10 | 5 | 13 | 58 | 99 |
| M50S25 | 12 | 6 | 16 | 69 | 121 |

Looking at the data, in half the time, twice the current is not delivered. For the 10-plate battery, 58 A is delivered over 8 h. But twice 58 A (116 A) is not delivered in 4 h. In 4 h, the internal losses are greater than the losses in 8 h. The faster the battery is discharged, the greater are the internal losses.

This is to be expected because losses are inherent. If the losses were of a fixed-resistive nature, then losses would be proportional to the square of the current. Losses are not of a fixed-resistive nature. A discharging battery has an internal capacity of chemical energy. As it discharges, some of that energy is converted to electric energy and some to heat. In Table 3.1, the converted portion is less than at 1.75 V/cell depth of discharge (DOD in the literature). Hence, the faster the battery is discharged, the lower is its efficiency.

In a fully charged battery, capacity is fixed, and during discharge, output and losses sum to capacity. Discharge efficiency is defined in the classical sense, as output/output + losses. If the discharge time is very long, with a small value of discharge current, losses are very small. If the current were about the same as the battery's trickle-charge current, the losses would be very small. Discharging for 120 h is a practical measurement of the capacity, because the losses approach zero at this rate of discharge.

In Table 3.2, the data are broken down to show the discharge efficiencies. The 120 h discharge is taken as capacity.

In charging the battery, much the same charge efficiency as discharge efficiency would result. Then if charge and discharge were comparable, of the energy delivered in 8 h of charging, about 75% would store in the battery. If this battery were to then discharge in 16 h at about 85% efficiency, only about 64% of the energy delivered to the battery would be usable.

**TABLE 3.2**

Battery in Table 3.1 Discharge Losses and Efficiencies

| Pos | Capacity A-h | | A-h at 40 h | | | A-h at 8 h | | | A-h at 4 h | | |
|------|-------|---------|-----|--------|---------|-----|--------|---------|-----|--------|---------|
| Plates | Total | Per Plt. | Out | Losses | Eff (%) | Out | Losses | Eff (%) | Out | Losses | Eff (%) |
| 10 | 600 | 60 | 520 | 80 | 87 | 462 | 138 | 77 | 396 | 204 | 66 |
| 12 | 720 | 60 | 640 | 80 | 89 | 554 | 166 | 77 | 482 | 248 | 67 |

## Battery Heat Flow

The battery losses are in amp-hours. For a 170 V battery using the 10 positive-plate cells, that is, 84 of these cells, a 4 h discharge yields 99 A for 4 h and a decrease in voltage from 187 to 153 V. The average voltage is 170 VDC, and the average load is 16.8 kW. In 4 h, the battery delivers 67.2 kW-h of electric energy. While discharging, the battery loses 34% in heat, or 52% of output in heat. This is an amount of 34.9 kW-h, or 119 thousand Btu. These losses are at an average rate of 8.73 kW of heat flow from the battery. The battery make-up is as follows:

### The Battery

Cell: 58 A at 8 h rate, 99 A for 4 h 1.8 V/cell discharge

Dimensions: 7.9 H, 5.9 W, 19.6 D

Module: 3 cells/module, 28 modules, 84 cells

Dimensions: 8 H, 18 W, 20 D

Weight: 225 lb

Stack: 10 modules, arranged in 3 stacks, 10, 10, and 8

Dimensions: 87 H, 20 W, 20 D

Weight: 2300 lb

Cabinet: 3 cabinets, each houses 1 stack

Dimensions: 96 H, 27 W, 24 D

Weight: 2450 lb (including stack)

### The Stack

Considering *only one stack* standing in the room, each stack has an envelope area of 87 by 80 in. of vertical area, or 6960 in.$^2$ Its top adds 400 in.$^2$, for a total outside area of 7360 in.$^2$, or 51.1 ft$^2$:

Areas of Stack

Outside area 51.1 ft$^2$ (4.75 m$^2$)

Air-wetted Area

Outside 51.1 ft$^2$

Inside 52.8 ft$^2$

In the stack, additional air-wetted areas of the tops and bottoms of the ten modules total 8000 in.$^2$ The total air-wetted area is 103.9 ft$^2$. The envelope area radiates heat to the room; the wetted area convects to the room. With 8.73 kW of heat flow generated in the battery, each cell generates 103.9 W, each module generates 311.8 W, and the stack generates 3.118 kW of heat flow, average during 4 h.

**Heat Flow Analysis of the Stack**

Heat flowing from the stack radiates and convects to the room. This assumes the stack alone standing in the room. Its surfaces are assumed isothermal, or are at an average temperature. Also, the stack is assumed as uniform in temperature throughout its mass at any one time. Its mass stores heat. The amount is referenced to a cold stack starting at room temperature. As the battery begins discharging and generating heat in its mass, the stack absorbs the heat and begins to heat up. After a time, it is fully hot and radiates and convects all of its generated heat to the room. The rate of its temperature rise above the room temperature is exponential. During its heating, the room temperature remains constant.

The heat stored in its mass is $q_m$. It is exponential, absorbing all of the generated heat at first, then decreasing exponentially to absorbing none when the battery has heated up. It absorbs heat only when the stack's temperature is increasing:

$$q_m = m c_p \, d\theta/dt$$

Here

     $q_m$ is in watts
     m is in pounds of weight
     $c_p$ is specific heat in watt-hours per lb degree Celsius
     $\theta$ is temperature rise in Celsius degrees
     t is in hours

The product, $m \, c_p$, is the thermal capacitance of a structure, in W-h/°C. Specific heat is relative to water's specific heat, 1.0 Btu/lb °F. Specific heat is a number less than 1.0 because most substances cannot store as much heat as water can.

Converting from Btu units, 1.0 Btu/lb °F is equal to 0.5274 W-h/lb °C.

At a heated-up temperature, when the stack is steady in temperature rise, the stack is radiating and convecting all of its generated heat to the room. The stack's radiated heat is

$$q_r = A_r \, F_a \, F_e \, \sigma \left[ T_s^4 - T_r^4 \right]$$

where

     $A_r$ is the radiating area in square meters
     $F_a$ is a factor from zero to one, which accounts for the radiating and receiving areas (one here)
     $F_e$ is a factor from zero to one, accounting for the emissivity of the radiating surfaces (0.85 here)
     $\sigma$ is Boltzmann's constant, $5.6 \times 10^{-8}$, for the units of this formula
     $T_s$ and $T_r$ are the temperatures of the stack and room in Kelvin degrees
     $\theta$ is temperature rise in Celsius degrees

The Kelvin-temperatures term is a polynomial. If expanded, and then the expansion is divided by $\theta$, the expression can then be the coefficient of $\theta$. Hence, this coefficient can then be evaluated for any temperature of the stack's surface. Let $T_r$ be 25°C + 273°C, or 298 K. Assume the hot surface is at about a 30°C rise above the room:

$T_s$ is 30 + 25 + 273, or 328 K. $T_r$ is 25 + 273, or 298°K

Then 5.6 $10^{-8}$ [(328)$^4$ – (298)$^4$]/$\theta$ = [5.6 (115.74 – 78.86)]/$\theta$.
Let this value be = 206.53/30 = 6.88, if $\theta$ were about 30°C:

$q_r = A_r$ (1.0)(0.85)(6.88) $\theta$

$q_r$ = 5.85 (A $\theta$) in units of W/m² °C, and with 4.75 m²

$q_r$ = 27.8 W/°C

for a surface temperature rise of about $\theta$ = 30°C rise.
   Its convected heat is

$q_c = h_c A_c \theta$

Here, $A_c$ is the area, in square feet, convecting heat. This area is the air-wetted area of the stack. This area is larger than the radiating area because the stack has module tops and bottoms, and the stack provides a half-inch air gap between the modules. This gap allows convection currents to flow. The coefficient hc represents the rate of convection in units of W/ft² °C. Its value is 0.36, representing an average for horizontal and vertical areas:

$q_c$ = (0.36)(103.9)$\theta$
$q_c$ = 37.4 $\theta$

The sum of the radiated and convected heat is

$q = q_r + q_c$ = (27.8 + 37.4)$\theta$
q = 65.2 W/°C for a temperature rise of about 30°C.

The heat generated in the stack, as stated earlier, is 3.118 kW. Then

$\theta$ = (3118 W)/(65.2 W/°C) = 48°C rise

This would be the rise after the stack had fully heated up.
   It would take about three time constants for the battery to fully heat up to a rise of 48°C. This would be about 9 h. In 4 h, the battery is still heating up when the load is disconnected. (If immediately charged with a heavy current, see following text about thermal runaway.)

## Exponential Temperature Response of the Stack

Together with the earlier equation, q = 65.2 W/°C, and the stack's thermal capacitance, the time constant for the stack can be found. The stack's thermal capacitance has units of W-h/°C. Dividing its thermal capacitance by 65.2 W/°C yields a value with a unit of hours.

What is needed is the specific heat of the stack, because its weight is known. The specific heat of these batteries and their structures have been determined.

## Battery Specific Heat

The specific heat of a pound of water is 1.000 Btu/lb °F. This is equal to 0.5274 W-h/lb °C. Specific heat is relative to a pound of water. As stated before, a valve-regulated lead–acid (VRLA) battery cell is dry inside its container. Each cell contains lead plates separated by electrolyte structures consisting of fiberglass mats soaked in the sulfuric acid electrolyte. Every other plate is connected to either the positive or the negative terminal of the cell. Two to six cells are packaged in a module. Modules of this type are usually stacked vertically to make up the battery. The terminals are on each module's side. A module's structure is suitable for stacking.

The stacks of battery modules may be modeled as an isothermal mass. Thereby, the overall specific heat is the weighed sum of its components' specific heats.

A range of one manufacturer's cells, ranging from 50 to 140 lb, resulted in a narrow set of values for cell specific-heat values. For each cell, summing the products of each component's weight times the component's specific heat, then dividing the sum by the cell's weight yielded values of cell specific heat of 0.217% ± 3.7% (Btu per pound °F). The values of specific heat that were used are listed in Table 3.3.

The weight of the module is the sum of the earlier materials' weights. The thermal capacity of the module is the sum of the products of component weight times its specific heat. The module specific heat is the thermal capacity divided by the module weight. The same is true of the cells.

The cell enclosure material for several manufacturers is ABS. Using the specific heat of ABS (polypropylene), 0.342, the following table results (weights are approximate).

**TABLE 3.3**

Battery Module Component Weights and Specific Heats

| Component | Weight | Specif. Ht. | Capacity |
|---|---|---|---|
| Steel | 2,655 | 0.121 | 322 |
| ABS | 4,425 | 0.342 | 1313 |
| Lead | 26,550 | 0.037 | 983 |
| Acid | 3,980 | 1.001 | 3984 |
| Fbr Gl. Mat | 6,640 | 0.342 | 2271 |

Specific heat is the sum of capacities/total weight: 9,073/44,250 lb = 0.21. The unit is Btu/lb °F. Multiply by 0.5274 to convert to W-h/lb °C. The weight of the stack is 2300 lb. Then

$$Cstack = (2300)(0.21)(0.5274) = 254.7\,W\text{-}h/°C$$

Let the time constant of the exponential temperature response be K. Then

$$K = 254.7/65.2 = 3.91\,h$$

Let the temperature rise be θ. Then

$$\theta = 48°C\left(1 - e^{-t/3.91h}\right)$$

Then with time of 4 h, θ = 48°C (1 – e^{-4/3.91}), θ = 48°C (0.640) = 31°C.

This is a stack temperature of 56°C in the room (133°F). Because its components begin to deteriorate at this temperature, this is a design limit for the battery. This is an average temperature based on a crude iso-thermal model. Some cell temperatures will be hotter than this representative value.

It would take about three time constants for the battery to fully heat up to a rise of 48°C. This would be about 12 h. In 4 h, the battery is still heating up when the load is disconnected. (If immediately charged with a heavy current, see following text about thermal runaway.)

## Heat Flow Analysis of the Stack in the Cabinet

The cabinet will impede the heat flow from the stack. It will be hotter than the room because the heat from the stack will flow through the cabinet, and the cabinet, to transfer the heat to the room, will have a temperature rise above the room temperature.

In Appendix A, the heat flow is derived for power cables in a conduit. Just as the cables generate heat, so does the stack. Just as the cables transfer their heat to the conduit and to the surroundings, so does the stack transfer its heat to the cabinet and to the room. Because of this similarity, the derived equations can be used for the stack.

Referring to Figure A.1, the thermal capacitance of the stack above becomes $C_1$. The thermal resistance $R_1$ is the thermal resistance from the stack to the cabinet, which can be found as the same as the resistance from the stack to the room, from q = 65.2 W/°C. Because resistance is °C/W, $R_1$ is 1/65.2, or 0.0153 °C/W. The thermal capacitance $C_2$ is that of the cabinet, and the thermal resistance from the cabinet to the room can be found from the dimensions of the cabinet. The weight of the cabinet is 150 lb. It is constructed of sheet steel and is painted inside and out. Sheet steel has a specific heat of 0.121, so its thermal capacitance is

$$C_2 = (150\,\text{lb})(0.121)(0.5274) = 9.6\,\text{W-h/°C}$$

The thermal resistance from the cabinet to the room can be calculated as was done for the stack's thermal conductance to the room (resistance is the reciprocal of conductance). The cabinet will not get as hot as the stack, say about 20°C rise above the room. Then the fourth-power Kelvin-temperature term is less:

$$\sigma\left[(20+25+273)^4 - (25+273)^4\right]/20°\text{C} = 5.6(102.3-78.9)/20 = 6.6$$

The outside area of the cabinet is from the dimensions: 96 H, 27 W, 24 D: 10,440 in.²

$A = 10,440$ in.² $= 72.5$ ft² $= 6.74$ m²

$q_r = A_r\, F_a\, F_e\, \sigma\, [T_s^4 - T_r^4] = (6.74)(1.0)(0.85)(6.6)(20°\text{C})$

$q_r = 756.2$ W for a 20°C rise

$q_r/20°\text{C} = 37.8$ W/°C

Convection heat flow is

$q_c = h_c\, A_c\, (20°\text{C})$ for a 20°C rise

$q_c/20° = (0.36)(72.5) = 26.1$ W/°C

The sum $(q_r/20°\text{C} = 37.8$ W/°C$) + (q_c/20°\text{C} = 26.1$ W/°C$) = 63.9$ W/°C

And again, referring to Figure A.1

$R_2 = 1/63.9 = 0.01565°\text{C/W}$

For the stack to the room, $q = 82.8$ W/°C for a temperature rise of about 30°C. Using this for the heat flow from the stack, which is nearly accurate, its thermal resistance value will be used for $R_1$.

$R_1 = 1/65.2$ W/°C $= 0.0153°\text{C/W}$

And

$R_1 = 0.0153°\text{C/W}$

$R_2 = 0.01565°\text{C/W}$

$C_1 = 254.7$ W-h/°C

$C_2 = 9.6$ W-h/°C

In Chapter 7, the results of Appendix A are further developed. The result is Equation 7.4.

The temperature rise of the stack inside the cabinet will be calculated from this equation, and the terms and relationships of the analysis in Appendix A will also be used. (This analysis is used in Chapter 7 for power cables, and also in Chapter 10 for a large motor.) Here, the heat generated in the stack is 3.118 kW for 4 h.

The results are as follows, calculating for the 4 h period:

Cabinet temperature rise above room is 19°C

Stack temperature rise above room is 38°C

Cabinet temperature is 44°C

Stack temperature is 63°C

This stack temperature compares with the stack temperature of 56°C if the stack is not inside the cabinet. If not inside the cabinet, the stack is 7°C cooler.

---

## A Different Use of This Battery

If another one of this battery were installed in a different plant, what performance could be expected? Let the new plant load discharge the battery in 2 h (to the same depth of discharge). Extending the battery characteristics in Tables 3.1 and 3.2 to a 2 h discharge requires knowing the discharge efficiency. Valve-regulated lead–acid batteries typically have a 54% discharge efficiency for 2 h of discharge to this depth of discharge. Using this for the battery, the value for amp-hours out is 324 A-h at a current of 162 A, with losses of 276 A-h in 2 h. (This plant has more load than the earlier plant, which draws 99 A of load.) The battery now delivers 27.5 kW to the load and loses 23.4 kW of heat in its cells. This is in 84 cells.

With 30 cells in a stack, the stack losses are 8.37 kW average for 2 h. This compares with the 3.118 kW for the 4 h discharge.

Now, using the same analysis for the 4 h discharge as used in "Heat Flow Analysis of the Stack in the Cabinet" section, for the 2 h discharge.

The results are as follows, calculating for the 2 h discharge:

Cabinet temperature rise above room is 34°C

Stack temperature rise above room is 59°C

Cabinet temperature is 59°C

Stack temperature is 84°C

It is strongly recommended that the stack not be installed in the cabinet.

## Charging the Battery and Thermal Runaway

If a large charging current were applied immediately to each of these two batteries to restore charge (the terminal voltages are low), the current would compare with the discharge current. The charging losses would compare with the discharge loss values also, and the batteries would continue to heat up, and overheat. The result would be that the battery could be damaged, or cause a fire.

To prevent thermal runaway, chargers are equipped with circuits that reduce charging currents if the battery is hot. Battery temperature of a central module is measured by a probe, consisting of a resistance temperature detector (an RTD) or a thermocouple (whichever the charger is equipped to use). Charging current is reduced by the charger, based on the measured temperature.

Required temperature measurement is not always understood by the user. The temperature probe is misinstalled. A probe's wiring between the charger and the module, or cabinet, is often not installed in a conduit, or the probe is attached to the frame (which is much cooler). Or, instead of two probes for the two chargers, only one probe is connected to one of the two chargers. How the probe tip is attached and the materials that are used are not well covered in the charger manufacturer's literature. This situation is avoided by more widespread knowledge of DC systems, loads, and batteries. And manufacturers must better describe installation requirements for the probes.

# Section II

# One-Line Designs

# 4

## The Largest Demand for 13.8 kV Switchgear

How much load can 13.8 kV switchgear supply? Medium-voltage switch-gear is rated from mainly 2.4 to 35 kV. Mostly, distribution is from 4.16 to 13.8 kV for industrial plants. Rarely does a plant distribution require higher than 13.8 kV, and for various reasons. The main reason is usually because the plant's size and load are within the demand limit shown here. There are three elements that bound the switchgear capacity limit:

1. The switchgear maximum bus current rating, 4000 A
2. The switchgear maximum interrupting rating, 1500 MVA
3. The large-motor current that feeds a system fault

The main bus maximum rating is 4000 A. Then the ampere limit of the switchgear is

$$\text{Load limit} = (4000\,\text{A})(13.8\,\text{kV})(1.732) = 95.6\,\text{MVA}$$

This would be the answer to the demand limit if the fault current associated with this load didn't exceed 62.8 kA. This value is the fault-current limit because of the maximum interrupting rating for medium-voltage switchgear, 1500 MVA, and with 13.8 kV:

$$\text{Fault-current limit} = (1500\,\text{MVA})/(13.8\,\text{kV})(1.732) = 62.8\,\text{kA}$$

There are two sources for switchgear fault current: the feeding transformer with its primary connected to the power grid, and the plant's large motors. Both sources' fault-current sum must not exceed 62.8 kA. Imagine a lineup of switchgear. The lineup has a main bus with a main breaker. This main breaker is connected to the supply transformer's secondary and supplies the switchgear's bus. From the bus, load breakers feed large motors and plant loads. The breakers feeding the large motors connect to transformers with secondaries rated for the motor voltage (see Figure 4.1). The load breakers feeding plant loads have to be rated to interrupt 62.8 kA or less. A plant load breaker having a three-phase fault on its feeder is fed fault current from the supply transformer and from the large motors.

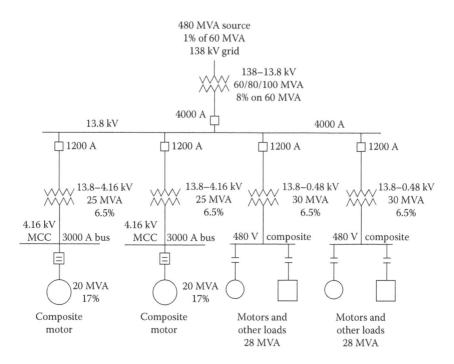

**FIGURE 4.1**
One-line sketch of a 96 MVA plant.

## Large-Motor Fault Current Contribution

When their terminal voltage is reduced to near zero because of a fault, the large motors generate a current larger than their normal current. The motors undergo a change in their magnetic fields and, with turning shafts, supply a fault current out of their terminals to their faulted source. This current is about six times its normal current, and decays exponentially to a negligible amount in about six cycles.

The decay time constant of a large motor, and that of the system, is the total effective time constant, $L/R = (\omega L/R)(1/\omega)$. Because system $\omega L/R$ is about 5, $L/R$ is in seconds, $\omega$ is 377, and system $L/R$ is about 13 milliseconds, or 1 cycle. Large motor $X''$ is about 17% (i.e., about six times, $X''$ being the subtransient reactance). Motor decay time constant, $T''$ (subtransient decay time constant), is about three cycles. The motor subtransient fault current, then, collapses in about nine cycles. (Collapse is essentially exponential. Collapse takes three time constants.) Because the system collapses faster than the large motors, and the fault voltage heads toward zero as the system collapses, the longer-collapsing motors finish last. The large-motor fault contribution is considered as six times the motor full-load current and is assumed as flowing during breaker interrupting time. This is a fair representation of the large motors of a plant.

The plant-load breakers interrupt more fault current than the main breaker and each large-motor breaker. Assuming a fault on the main bus, the main breaker has to interrupt only the fault current from the supply transformer. The motor-feeder breakers feed the bus fault also, but the motor-feeder currents flow only through their breakers to the fault. With a fault on a large-motor feeder, the main breaker feeds the fault current from the supply transformer through the faulted large-motor breaker. The other large motors feed their fault currents through the faulted large-motor breaker. This fault current, the sum of the main breaker current and the other large-motor fault currents, is less than the fault current with a bus fault, which has the full large-motor fault current. With a fault on a plant-load feeder, the main-breaker fault current and the full large-motor fault current flow through the faulted feeder's breaker.

### Large Motors

Large-motor voltages are rated 2300 and 4000 V. A rating of 1330 V is rarely found. Although large motors can be as large as 6500 Hp, driven by an 800 A contactor, this size is also rarely found. Most contactors in a medium-voltage motor control center are 400 A rated and can supply up to a 3000 Hp motor. Motors come in several varieties, squirrel-cage and synchronous, capable of reversing and not reversing, and with reduced voltage starting. Large cooling towers can have fan motors that reverse and have a reduced speed. This operation minimizes ice buildup in winter but is obsolescent. Cooling tower fans rarely exceed 200 Hp, and more and more rely on variable-speed drives.

## Small Motor Contribution

The plant-load feeder has negligible contribution of fault current during a system fault. Why does the plant-load feeder have negligible feed of fault current? The plant load consists of mostly motors, but these motors' voltages are low-voltage values. A composite of all these motors would exhibit the same X″ value of about 17% as for the large motors, but the subtransient time constant is much shorter.

Where windings in large motors are for high voltages (many turns) and low currents, windings in small motors are for low voltages (600 V) and high currents. This makes their subtransient time constants much shorter. The T″ values are about 0.01 s. The fault current, then, is gone in 0.03 s, or less than two cycles, with most of the current decaying in the first half cycle.

*Because protective devices require more than one cycle to sense a fault current, a fault current must last more than a cycle. With most of the small-motors' fault current decaying in the first half cycle, the fault current contribution of low-voltage motors is commonly not counted in fault analysis. Also, variable-frequency drives feeding motors are not a source of fault current.*

---

## Medium-Voltage Motor Control Centers

Medium-voltage motor control centers are built with motor controllers in a two- or three-high arrangement. Each controller has a disconnect switch, fuses, a contactor, and a control. A 1500 Hp fan motor might require five contactor compartments, one each for forward, reverse, and reduced voltage starting and two more if the motor winding is for two speeds. This motor, with all these duties, would require three sections of motor control center, each section 36 in. wide and about 8 ft high.

A motor controller consists of a knife switch that disconnects fuses, which in turn feed a contactor, usually rated 380 or 400 continuous amperes. A control-power transformer connects downstream from the fuses. Its secondary is rectified, making 240 VDC for control. Motor controllers are usually stacked and connected to a top-mounted main horizontal bus. Other components can be in controllers, such as an inductor for reducing voltage during starting.

Motor control centers come with main buses having a maximum rating of 3000 continuous amperes. This limits the motor control center's load and the transformer size feeding the motor control center. This also limits the main bus's feeder and its breaker size. A motor control center's maximum load is

$$(4.16\,\text{kV})(1.732)(3\,\text{kA}) = 21.6\,\text{MVA}$$

The feeding transformer's primary current is

$$(3\,\text{kA})(4.16\,\text{kV}/13.8\,\text{kV}) = 904\,\text{A}$$

Then the 13.8 kV bus breaker will be rated 1200 A.
And the transformer size can't exceed

$$(1200\,\text{A})(13.8\,\text{kV})(1.732) = 28.7\,\text{MVA}$$

This size exceeds a unit substation transformer's size, which can go to 25 MVA. This 28.7 MVA is the low end of large transformers' sizes. (See IEEE C57.) The transformer will be rated 25 MVA. Probably, there will be a small expedient load on the motor control center, say 2 MVA, leaving the *maximum* motor load of 20 MVA on this motor control center. This total load is limited to 22 MVA so that the 3000 A bus capacity isn't exceeded.

This 25 MVA transformer rating for the 3000 A motor control center provides a margin for the motors of 25 – 20 = 5 MVA, or 5/20 = 25% load margin on this transformer that is supplying the motor control center. Some margin is needed to allow for voltage regulation during motor starting, and 25% is ample.

If the motors have variable-speed drives, motor starting inrush current doesn't occur, and transformer margin isn't required. Also, drives do not contribute fault current. If one finds more 4160 V motor load than 20 MVA in the plant, then one will provide more motor control centers, transformers, and medium-voltage bus breakers.

## Maximum Plant Motor Load

If one makes a list over the years that breaks down industrial plant electrical loads, one finds a fairly stable ratio, shown in the following table. This ratio exists over a wide range of plants that make all kinds of industrial products in many industries. The table also applies to commercial skyscrapers. Of course, there are exceptions. The makeup is listed in Table 4.1.

Usually, steam heats the process, and the pumps may be turbine or electrically driven. Efficiency imposes use of steam to power the process, rather than using plant steam to make electricity, or buying the electricity, then using the electricity to power the process. Using steam to power the process lowers the electric bill. Lately, for variable speed requirements, motors with variable-frequency drives give better efficiency and performance.

With half the demand taken by large motors, for our 96 MVA maximum cap, 48 MVA would be estimated for large motors. We now can count the number of medium-voltage motor control centers. With a limit of 20 MVA per center, three are required for this maximum-demand plant. But if we make a small simplification, with the large-motor load estimate trimmed from 48 to 40 MVA, we can require only two. As seen in the following, we need simplification, and we can justify this small loss of estimate. More on this later.

### TABLE 4.1

Universal Common Load Makeup

| Load Type | Percent of Total | Notes |
|---|---|---|
| Large motors | 50 | Medium voltage, 300 Hp and up |
| Small motors | 25 | Low voltage, 3–250 Hp |
| Lighting | 15 | Low voltage |
| Miscellaneous | 10 | Nonprocess building heating, etc. |

## Fault Analysis of the Maximum Plant

Our analysis will consider only the reactances of the network, because including the resistances would decrease the results by a negligible 1% or 2%. Also feeders in our network are neglected because they are usually short in industrial plants. Usually, per unit methods are used to analyze networks. This makes the work easier if there are several voltage levels involved. I'm using plain old Ohm's law because only 13.8 kV is involved. Also, computers don't help here. The manual method you see herein (with thorough checking) is free of garbage in, garbage out.

A word about computing accuracy. If you have a tested transformer, and the tests show that the actual capacity is 60.3 kVA, you only know the capacity to an accuracy of three significant figures. But you did your analysis months before the test, and you could only estimate the test results when you did your calculations. It is silly to use a value of 60.00 kVA (or more digits) in your calculations and then believe you have accuracy to four significant figures. You don't. You have accuracy of only two significant figures. I'm calculating using three, or four, space figures, but only the first two mean anything.

Figure 4.1 shows the medium-voltage one-line sketch of the plant as discussed herein. Our estimated 40 MVA of large-motor load shows as two circuit breakers, transformers, and motor control centers, each with fully loaded 3000 A main buses. These are composite loads. More motor control centers and their transformers are possible, but their loads will not exceed the sum of these two loads. The low-voltage load, estimated at 56 MVA, also shows up as two circuit-breaker circuits, but these also are composite loads. As discussed earlier, a 1200 A circuit breaker can supply 28.7 MVA, and two can supply 57.4 MVA. Whether there are more than two feeders for plant low-voltage load doesn't affect the short-circuit analysis because plant low-voltage load breakers don't contribute fault current. Also as discussed earlier, the maximum fault current flows through either of these two plant circuit breakers. A single medium-voltage motor represents the motor load on each medium-voltage motor control center. This also is a composite motor. Because all actual motors are modeled as having an X" value of 17% and a T" value of 3 cycles (0.05 s), these are the values for each of the two composite 20 MVA motors.

The transformers feeding the medium-voltage motor control centers are rated 25 MVA. They have a nominal impedance of 6.5%. This is the ANSI/IEEE standard recommended value.

The main transformer connects to the high-voltage grid. The transformer's impedance is 8% and is based on the transformer windings having breakdown insulation levels corresponding to recommended values for high and medium voltages. This 8% value also is the ANSI/IEEE standard recommended value.

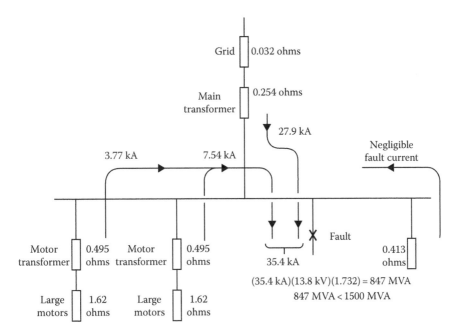

**FIGURE 4.2**
Impedance network sketch of a 96 MVA plant.

The grid Thevenin impedance assumes that the equivalent grid MVA is a size of eight times that of the main transformer's base rating, that is, without any forced cooling (ONAN rating: see standards). So long as this ratio is high, the grid MVA impedance has little effect on the value of fault current that flows through the main breaker. The smaller this ratio, the less the fault current. The Thevenin impedance is then one-eighth of 8%, or 1%.

Figure 4.2 shows the impedance network. All values are referred to the secondary 13.8 kV bus. For example, the main transformer and grid together are 9.0%.

This 9% impedance value refers to the ONAN rating. This size transformer would commonly have two stages of forced air cooling and be rated at 96 MVA with both cooling stages operating. Following transformer standards, the transformer would have a 33% capacity for each stage of cooling. And *without* the cooling stages operating (the ONAN rating), the transformer rating would be (96 MVA/1.67), or 57.5 MVA, or more likely 60 MVA:

$$60 \text{ MVA}/(13.8 \text{ kV})(1.732) = 2510 \text{ A for the ONAN secondary rating}$$

ONAN amperes on the 13.8 kV bus is $A_{ONAN} = 2510$ A, and transformer bolted fault current is, with 9.0% impedance, 2510/0.09 = 27.9 kA.

$$V_\Phi = 13.8 \text{ kV}/1.732, \text{ or } 7.97 \text{ kV}$$

Then

$$Z_{tot} = 7.97 \, kV/27.9 \, kA = 0.286 \, ohms \, (\text{which includes the grid contribution})$$

In the same manner, the other impedance values are calculated. Figure 4.2 shows these values. The motors are calculated as 20 MVA on a 13.8 kV base. Their impedances have a 17% impedance value on the 13.8 kV base: 20 MVA/(13.8 kV) (1.732) = 837 A, and (7. 97 kV/0.837 kA)(0.17) has a value of 1.62 ohms.

Figure 4.2 shows the grid and main transformer fault current as 27.9 kA, the motor contribution as 7.54 kA, and their sum as 35.4 kA. Also, the product of the fault current, voltage, and square root of three calculates to 847 MVA interrupting duty. We needed to show that the fault duty was less than 1500 MVA, which it is: only some 56% of the limit. When we shaved the motor load from 48 to 40 MVA, we didn't shave much. The motors contribute about 22% of the total fault current. Increasing this contribution up a little wouldn't increase the fault current much. And similarly, making the grid more powerful than 480 MVA wouldn't increase the current much through the main transformer's path. Making the grid weaker would only lower the fault current. Increasing the available fault current from the grid, and its effect, is discussed in Chapter 5.

---

## Conclusions

The maximum load that medium-voltage switchgear can feed at 13.8 kV is 96 MVA. The bus limit of 4000 A defines this value. If the voltage were the uncommon value of 15.0 kV, the interrupting MVA would increase to about 1020 MVA, still within the 1500 MVA limit.

> The most common large medium-voltage switchgear is with two equal main transformers, and their secondaries connected in a main-tie-main arrangement in the switchgear. These two sources cannot ever be connected together because the available fault current would exceed the switchgear's interrupting capability. Therefore, the mains and tie breakers are heavily interlocked to prevent both the mains and the tie to be closed at the same time. This main-tie-main arrangement is fully redundant.

A plant, or one site, needing more than 96 MVA of power is uncommon. Although a power-generating plant can have a large transformer to step up its generator output, rated 250 MVA or more, the plant's auxiliary load is only about 10% of the output, or about 25 MVA.

If a plant load exceeds 96 MVA, the switchgears can be increased but will require a third main source: a main-tie-main-tie-main arrangement. More than likely, the loads will be spread out and be better fed from multiple distribution sources.

## Exceptions

The U.S. Gulf Coast has the greatest concentration of plants, probably in the world. The plants are mainly refineries and petrochemical plants. Years ago, when the Houston utility was known as Houston Light and Power, because of this concentration, it was the largest utility in the United States. Their sales department was asked if they would list some of their largest customer demands. The largest was about 60 MVA, and the smallest was about 10 MVA in their list.

One exceptionally large plant was a pulp mill in Alabama. The usual way to make pulp is to chip logs and digest the chips into cellulose-fiber pulp. This plant ground the logs into pulp and separated out the turpentine. It had three grinder trains, each with huge 12.7 kV synchronous motors. Each train drew 25 MW. (With synchronous motors, you can control the vars by controlling the motors' excitations, and have unity power factor.) Alabama Power designed the service connection and chose four 30 MVA transformers feeding two switchgear lineups. Each lineup was a main-tie-main arrangement.

Another example is an aluminum smelter. Here, the demand can easily exceed 96 MVA. There are no large motors: the auxiliary load is about 10% of the demand. The basic load is paralleled potlines, wherein a potline is composed of series-connected reducing cells. Thousands of DC amperes flow through the cells, converting alumina powder to molten aluminum. Each potline has rectifiers feeding it. Each cell in the potline drops a voltage of about 1.5 V, with maybe 200 cells in a line. The rectifiers deliver 320 VDC at 200,000 A, or 64 MVA to each potline. A plant like this could have six potlines and a demand of 430 MVA. Harmonic suppression is mainly done in the transformers feeding the potline's rectifier sections. The windings are sectioned and connected to shift phases every 15°.

# 5

## The Minor Influence of the Grid Thevenin Impedance on Plant Faults

In the example of Chapter 4, a limiting service and switchgear size of 96 MVA is demonstrated. A value for the grid of 480 MVA available fault current was used, making a Thevenin impedance of 0.032 ohms, one-eighth that of the main transformer's impedance. This 480 MVA value is eight times the ONAN rating of the main transformer. Using the ONAN rating as a reference value of 60 MVA, the grid MVA can range from about 3 times 60 MVA to well over 20 times 60 MVA. Given a main transformer, the grid contribution can be addressed as a multiple of the ONAN rating of the service transformer. With three times, the source is a very weak grid. This could be found with a long pole line extension to the main transformer.

With 20 times, the grid is very strong and more representative. In a loop switchyard, with a design value of 3000 A in its switchyard bays, the bay's conductors, buses, and breakers are near their limits. And with 138 kV incoming lines, the driving MVA is a value of 717 MVA. This is about 20 kA of fault current at 138 kV in the yard. This is also 12 times a 60 MVA transformer's ONAN rating.

Let's see how the fault current through a main transformer changes with a range of source impedances. Let's use a ratio of the transformer impedance for the comparison. The weakest source will be 3 times, and the strongest will be 20 times. The weakest impedance will be one-third, and the strongest impedance will be one-twentieth. Further, we will place the actual impedance at *one-sixth*. We will use a transformer with an ONAN rating of 60 MVA feeding 13.8 kV plant distribution (phase voltage of 7.97 kV). The transformer has an *impedance of 0.254 ohms*. At one-sixth that of the transformer, the grid Thevenin impedance is 0.0423 ohms. The sum of the source's and transformer's impedances is $1 + 1/6$, or 7/6th times 0.254, or 0.296 ohms. Thus, the actual fault current is

$$I_f = 7.97 \, \text{kV}/0.296 \, \text{ohms} = 26.9 \, \text{kA actual fault current}$$

Now, assume that the source's capacity is at three times the transformer's 60 MVA. The fault current would be 23.5 kA at the transformer's secondary terminals. And with the source's capacity of 20 times, the fault current would be 29.9 kA. These values vary from 26.9 kA ± 12%. But with a better estimate for the grid's Thevenin impedance, actual error can be much less with just knowing a small amount about the plant's source.

If your plant is at the end of a 110 kV long line, your estimate can be around four or five times. And if your plant is fed from a loop substation at 345 kV, your estimate can be around 20 or 40 times. This is typical if your main transformer is about 20 MVA. At smaller sizes, your plant can be fed from medium-voltage sources. Still, a multiple of your main transformer's rating is a useful tool.

Hence with this spread, from 4 to 40 times, the error in the fault-current supply is ±12%. The error from applying a basic knowledge of the supply to add to the transformer's impedance can reduce the fault-current error to a negligible 5% or less.

For example, let the ONAN 60 MVA transformer be supplied from a source that is eight times the transformer's rating. Then the available source is 480 MVA, its impedance is 0.032 ohms, and the transformer impedance is 0.254 ohms. (These values are shown in Figure 4.2.) With a 13.8 kV secondary

$$\text{Actual fault current} = 7.97\,\text{kV}/(0.032+0.254) = 27.9\,\text{kA at }13.8\,\text{kV}$$

But let the estimated source be 960 MVA, or 16 times, and the source impedance is estimated as 0.016 ohms:

$$\text{Estimated fault current} = 7.97\,\text{kV}/(0.016+0.254) = 29.5\,\text{kA at }13.8\,\text{kV}$$

The fault current increases from 27.9 to 29.5 kA. This estimated fault current is only 5.7% more than the actual fault current, an estimate change from the actual 8 times 60 MVA to 16 times, or double the system available capacity. This is changing from a moderately sized switchyard supplying the plant to a very large switchyard.

The 5.7% error here is on the conservative side, and it is compared to the tolerance of the transformer's impedance. Most suppliers can provide a tolerance of 4%, although the ANSI IEEE C57 standard allows 7.5%.

It is better to overestimate the source than to underestimate the source. With an overestimate, the results are on the conservative high side.

# 6

## Use of a Three-Winding Step-Up Transformer

### Introduction

Years ago, utilities generated, transmitted, distributed, and sold electricity. They were regulated by the states and had their own operating territories. Recently, they were disassembled into separate corporations that are, and are not, regulated. Also recently, climate change created a need to reduce carbon dioxide emissions. Now, the federal government tax-subsidizes renewable generation, and we hope this source will eventually reduce its costs of delivering a kilowatt-hour to consumers. But there is a hitch.

Renewable sources are not reliable. They are not capable of supplying an electric load 24 h/day. To do so, a storage means must come into use for when the sun doesn't shine and the wind doesn't blow. Many storage means are possible (molten salt, fly wheels, pumped storage into a high lake, batteries) but so far these devices are too costly to design, produce, and operate. And they have been under development for many decades.

Any storage means has an inability to deliver all of the energy sent to it. This is inherent. Its charging and discharging losses cause the reduction. With storage batteries, a typical value can be that, of the charge energy, half can be delivered on discharge, with the other half dissipated in losses.

Another huge requirement is from a turnover in cars and trucks. New vehicles are to be battery-electric powered.* As electric vehicles become more numerous, more electric generation and distribution will be demanded to supply them. Electricity has been around since about 1890. Since then, growth has doubled every 3 years, and utilities have had to keep doubling their assets since then to meet this demand. With electric vehicles,

---

* Hydrogen fuel cells to power vehicles have a huge hitch. Unlike natural gas, hydrogen can't be mined. Hydrogen is mainly made from natural gas. The world presently *consumes* about 50 million tons of hydrogen per year. Most of this is made and used in refining oil, breaking down heavy hydrocarbons to lighter hydrocarbons by hydrogen cracking. Hydrogen made from natural gas is the cheapest present source and produces less pollution than if made from oil or coal. In making their intermediate hydrogen fuel, fuel-cell-driven cars will typically consume natural gas and electricity. These cars will thus pollute from making their fuel.

this doubling will continue. Instead of our present gasoline refineries, electric generating stations will replace them (hopefully renewable). And more transmission and distribution will continue to be needed also.

The future doesn't promise any reduced cost of backup storage.

Instead of storage devices, renewable-source systems, if they are not to be adjunct to fossil utilities, must probably use methane or other abundant gaseous low-emission fuel. And here, the generation will probably be by the combustion turbine. This device is fairly cheap to build and operate. And to be flexible in its use, it is suited to on–off daily cycling. Alternatively, diesel generation with gaseous fuel is also viable.

The combustion turbine generator has a fairly stable cost per kilowatt for both small and large units. However, if its role in reliable renewable generation is to back up wind and solar generators, several small-sized units are best, rather than a few large units. The small units can be more quickly started and can best adjust to small step changes in demand.

## The Small Off-Power Generating Unit

Within each small unit, its auxiliary electrical power can be supplied from two sources: from the generator, and from the high-voltage system receiving the generated power. This high-voltage system is used for starting the small unit. A more redundant distribution beyond these two sources offers no advantage in reliability and would cost more.

Therefore, a most economical arrangement is the normal auxiliary power supplied from a third winding in the step-up transformer. The second starting source from the high-voltage system serves as the backup source. This arrangement is shown in Figure 6.1. This generator size is typical and not optimum. The optimum size depends on the size of the renewable plant and its power grid.

## The Small-Unit Auxiliary Load

Figure 6.1 shows the single-line diagram of the small unit and its three-winding transformer. Here, the generator is a 17 MW 0.85 power factor size tied into a 138 kV switchyard. This is a typical application. The generator breaker is the tie breaker in the switchyard. A 1000 A bus connects the generator's terminals to its transformer terminals.

The transformer's third winding is shown with a 2 MVA rating at 4160 V. A winding ratio of 33 to 1 for this auxiliary winding keeps the transformer's

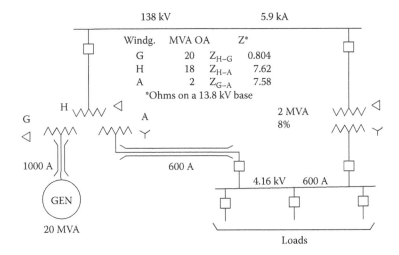

**FIGURE 6.1**
One-line diagram of a small off-power generating unit.

windings typical. The 2 MVA rating is 10% of the generator output. This also is a conservative estimate of the turbine unit's auxiliary load. The switchgear main breaker is tied to the transformer's three terminals with a 600 A bus. This bus is in the protective zone of the transformer.

A breaker at the transformer would have no purpose. The transformer protective zone extends from generator to switchyard breaker to auxiliary switchgear main breaker. Faults in this zone are very rare. A fault in this zone trips the generator (excitation is turned off and fuel to the turbine is shut off): loss of generation trips the generator's breaker, trips the main breaker to the auxiliary bus to isolate it from the transformer, and transfers the bus, with its unneeded, soon-to-be-shut-down load, to the start-up source. Another small unit comes online to replace lost generation from this unit.

By simplifying connections in this scheme, it saves transformer core iron and eliminates large breakers. This scheme thus costs less than alternatives. Adding auxiliary transformers and breakers does nothing for reliability.

I once was designing a small combustion turbine plant. My auxiliary load analysis didn't conform to the breakdown shown in Chapter 4 for industrial load constituents. A large motor load that took up about 50% of the total was missing. The auxiliary load was small motors, lighting, and other small loads. I was beginning to think that such generating units were an exception to the universal breakdown in Chapter 4. Then the gas supplier said she had no medium-pressure distribution in the local region and couldn't supply the fuel at the pressure demanded by the combustion turbine. Her 20 lb supply pressure couldn't meet the 75 lb that the turbine's manifold needed. So a compressor was added to the auxiliary load, and the compressor almost doubled my load analysis total.

## The Transformer

After factory testing, only impedance measurements between windings are possible. As shown in Figure 6.1, the high-voltage winding connecting to the 138 kV switchyard is rated 18 MVA. The generator winding is rated 13.8 kV at 20 MVA, and the auxiliary winding is rated 4.16 kV at 2 MVA.

These MVA values are derived from the generator size and size of its auxiliary load, 10% of the generator rating. The high voltage is based on the transmission region's local 138 kV system. The generator voltage is a choice for economical generator design. The auxiliary voltage is a standard voltage for a switchgear load of 2 MVA. Also, there are transformer standard values for winding insulations (BILs) and bushing designs.

To provide the generator with ground fault protection, its transformer winding is delta connected, and the generator's windings are connected wye with neutral protection. To provide the auxiliary bus with neutral protection, its winding is wye connected, with a neutral grounding resistor. The high-voltage winding is connected as required by the transmission line's protection requirements, usually delta as shown. This is dictated by the transmission company. At least one of the three windings is delta connected to provide third harmonic suppression.

If the auxiliary winding and the high-voltage winding are both wye, then the start-up transformer must also be such that its secondary winding is in phase with the auxiliary winding. Both bus sources must be in phase for rapid bus transfer between sources. With the start-up's high-voltage delta, the start-up transformer's secondary must be zigzag to match the wye-wye windings of the main transformer.

Ratings of the windings are not critical. Economic output requires that the transformer not limit the generator's capability. The high-voltage winding rating, less than the generator's rating, need not be larger if the actual auxiliary load turns out to be less than its winding's capacity. This is not a concern for two reasons.

The first is that the windings' ratings are based on a 30°C average daily temperature and an oil temperature of a 55°C rise. A 30°C average day is an extremely warm day, and the transformer has a capability of a 65°C rise. The second reason is that, although as turbine inlet temperature cools, the turbine output capability increases; the transformer ratings also increase as the outside air cools. This increase in a transformer's capability is more than the increase in its turbine's capability.

The factory values on the transformer's nameplate *could be* as follows:

| | |
|---|---|
| Hi-V winding to gen. winding | 8% of 18 MVA, 138 kV base |
| Hi-V winding to aux. winding | 9% of 18 MVA, 138 kV base |
| Gen. winding to aux. winding | 5% of 20 MVA, 13.8 kV base |

These measurements are each made with the uninvolved winding open. That is, the high-voltage-to-generator impedance is measured with the auxiliary windings unconnected.

But these impedances must be specified before the transformer is manufactured, and the transformer being measured has to be designed and built to meet the specified winding MVA ratings, the winding connections, and the impedances (along with other requirements). Then the impedances must first be designed by analysis.

Analysis requires a transformer wye impedance model, as shown in Figure 6.2. Also, the impedances have to be of a common base. Thus, the common-voltage impedance values must be developed. Once developed, the following equations convert the nameplate-converted common-voltage values to wye values. Figure 6.2 is the impedance diagram for analysis. All resistive values are neglected.

$$Z_G = \frac{1}{2}\left(Z_{H-G} + Z_{G-A} - Z_{H-A}\right)$$

$$Z_H = \frac{1}{2}\left(Z_{H-G} + Z_{H-A} - Z_{G-A}\right)$$

$$Z_A = \frac{1}{2}\left(Z_{H-A} + Z_{G-A} - Z_{H-G}\right)$$

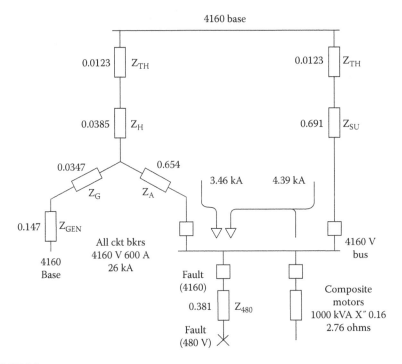

**FIGURE 6.2**
The impedance diagram of the unit with 4160 V base.

These three equations are derived from the way the transformer is measured—that is, from $Z_{H-A}$ being measured with the generator terminals unconnected, and so forth.

## Specifying the Transformer

An 18 MVA *two-winding* transformer with voltages of 138–13.8 kV would have an 8% typical impedance (see recommended values in ANSI IEEE C57). Measured on the 13.8 kV side, this would be a value of 0.846 ohms, as calculated in the following. Looking at the transformer's three-winding core, it would have on each of its legs comparable windings found on this two-winding 18 MVA transformer plus the third winding. Its core would be a little taller than the two-winding transformer's core, to allow the third winding to be fitted.

## Fault Current in the Auxiliary Distribution

The analysis starts with the limit of the short-circuit current rating of the 480 V distribution to small loads. If a feeder to a 480 V load were to fault, its circuit breaker at the feeding 480 V bus would interrupt the fault current. Most 480 V distribution equipment is rated to interrupt no more than *65 kA of fault current*. Next, this 480 V bus would have a transformer feeding it from the 4160 V auxiliary bus. The largest transformer size feeding any plant 480 V bus would be about 2500 kVA. This is based on most 480 V switchboards and switchgear, which have a maximum bus rating of 3200 A:

$$kVA = (3200\,A)(0.480\,kV)(1.732) = 2660\,kVA$$

The nearest standard transformer size is 2500 kVA.

Even though *this exceeds the auxiliary load of 2000 kVA*, it is the extreme value, which is of interest here.

The standard minimum impedance for this transformer is 5.5%. Then the transformer's impedance in ohms is

$$I_{F\,Load} = 2500\,kVA / \left[(0.480\,kV)(1.732)\right] = 3007\,A\ \text{and}$$

$$Z_{480} = 0.055\left[(480)/(1.732)(3007\,A)\right] = 0.00507\ \text{ohms on 480 V}$$

Converting to 4.16 kV:

$$Z_{480} = (4160 \text{ V}/480 \text{ V})^2 \, 0.00507 = 0.381 \text{ ohms on } 4160 \text{ V}$$

If the fault happens when the auxiliary bus is fed from the 2000 kVA start-up transformer, the start-up transformer is in series with the 2500 kVA transformer feeding the fault. The impedance of this transformer is 8%:

$$(I_{Fl \, Load}) = 2000 \text{ kVA}/(4.160 \text{ kV})(1.732) = 278 \text{ A and}$$

$$Z_{SU} = 0.08 \left[ (4160)/(1.732) \, 278 \right] = 0.691 \text{ ohms on } 4160 \text{ V}$$

$$Z_{SU} = (480/4160)^2 \, (0.691) = 0.0092 \text{ ohms on } 480 \text{ V}$$

The two transformers add to $(0.0092 + 0.00507) = 0.0143$ ohms. Then the fault current would be

$$I_{FAULT} = 480/(1.732)(0.0143) = 19.38 \text{ kA} < 65 \text{ kA}$$

Looking at Figure 6.2, the most fault current that can flow would be from the composite motor in parallel with the start-up transformer. This added source will increase the 19.4 kA fault current.

If both main bus breakers are closed when the fault happens, then the fault current flowing through $Z_A$ from the three-winding transformer will contribute even more current to the fault. The most fault current that can flow is from the transformer, and from the start-up transformer in parallel with the composite motor. Both sources would be in parallel with each other. If $Z_{NET}$ is this paralleled minimum impedance, it would be in series with $Z_{480}$, which is feeding the fault, with its value of 0.381 ohms (on a 4160 V base). Then

$$I_{FAULT} = 2400 \text{ V}/(Z_{NET} + Z_{480}) = 2400 \text{ V}/(Z_{NET} + 0.381)$$

If the fault current were 65 kA, on the 480 V base, it would be $480/4160(65 \text{ kA}) = 7.5$ kA on the 4.16 kV base.

$7.5 \text{ kA} = 2400 \text{ V}/(Z_{NET} + Z_{480})$ is the limiting impedance imposed by 7.5 kA.

$$(Z_{NET} + 0.381) = 2400 \text{ V}/7.5 \text{ kA} = 0.32 \text{ ohms on } 4160 \text{ V base.}$$

But $Z_{480}$ is already bigger than this limiting impedance; therefore, the fault current cannot exceed the 65 kA limit, and the three-winding transformer's impedances cannot be set based on fault current in the 480 V auxiliary distribution.

Unlike larger generating units, this generating unit is too small for fault-current limits to determine winding impedance requirements. The total fault current from the start-up transformer, large motors, and the three-winding transformer must also be evaluated, as done in the following.

---

## Transformer Impedance Values

As discussed in "Specifying the Transformer" section, the windings on a core of an 18 MVA two-winding transformer would have, on a *13.8 kV base* and 8% impedance, an impedance of

$$I_{rated} = 18,000 \, kVA/(1.732)13.8 \, kV = 753 \, A$$

$$Z = 0.08 \, (Z_{rated}) = 0.08 \left[ (13.8 \, kV)/1.732(753 \, A) \right] = 0.08 \, (10.57 \, rated) = 0.846 \, ohms$$

An impedance similar to the start-up transformer, having 8% impedance also, but on a 13.8 kV base, is

$$I_{rated} = 2000 \, kVA/(1.732)13.8 \, kV = 83.7 \, A$$

$$Z = 0.08 \, (Z_{rated}) = 0.08 \left[ (13.8 \, kV)/1.732 \, (83.7 \, A) \right] = 0.08 \, (95.2 \, rated)$$

$$= 7.62 \, ohms \, on \, a \, 13.8 \, kV \, base$$

Because the generator winding may be suitable with leakage for an 8% base, as with the start-up transformer, this percentage will be used instead of a 6.5% typical value for a 13.8–4.16 kV transformer (see IEEE C57).

With these impedances, the leakage fluxes giving these values should also be similar. Then for $Z_{H-G}$ on each leg of the core, the two windings should have comparable leakage impedance. If they were equal, their value would be half of 0.846, or 0.423 ohms.

This is suitable for the high-voltage winding with its 18 MVA rating. However, the generator winding is rated 20 MVA. Adjusting for its rating, $18/20(0.423) = 0.381$. Adding this to 0.423 for the high-voltage winding, $(0.381 + 0.423) = 0.804$. This is a suitable value for $Z_{H-G}$.

For $Z_{H-A}$, the H winding impedance would have the value of 0.423 ohms, and similar to the start-up transformer, with a total of 7.62 ohms for $Z_{H-A}$, the smaller winding would have most of the leakage impedance. Subtracting 0.423 from 7.62, the impedance of the A winding would be 7.20 ohms on a 13.8 kV base.

Then for $Z_{G-A}$, the G winding, of 0.381 ohms, and the A winding, of 7.20 ohms, would equal the $Z_{H-A}$ value: $(0.381 + 7.20) = 7.58$ ohms for $Z_{G-A}$, on a 13.8 kV base.

## Calculating Wye Values

The following summarizes the earlier values, all on a 13.8 kV base:

$$Z_{H-G} = Z_H + Z_G = 0.804$$

$$Z_{H-A} = Z_H + Z_A = 7.62$$

$$Z_{G-A} = Z_G + Z_A = 7.58$$

Substituting these values in the measured-to-wye model equations on a base of 13.8 kV

$$Z_G = \tfrac{1}{2}\left(Z_{H-G} + Z_{G-A} - Z_{H-A}\right)$$

$$Z_H = \tfrac{1}{2}\left(Z_{H-G} + Z_{H-A} - Z_{G-A}\right)$$

$$Z_A = \tfrac{1}{2}\left(Z_{H-A} + Z_{G-A} - Z_{H-G}\right)$$

$$Z_G = \tfrac{1}{2}\left(0.804 + 7.58 - 7.62\right) = 0.382 \text{ ohms}$$

$$Z_H = \tfrac{1}{2}\left(0.804 + 7.62 - 7.58\right) = 0.423 \text{ ohms}$$

$$Z_A = \tfrac{1}{2}\left(7.62 + 7.58 - 0.804\right) = 7.20 \text{ ohms}$$

For the circuit in Figure 6.2, the analysis is on a base of 4.16 kV. The following convert to the analysis base of 4.16 kV:

$$\left(4.16/13.8\right)^2 = \left(0.0909\right)$$

$$Z_G = \left(0.0909\right)0.382 \text{ ohms} = 0.0347$$

$$Z_H = (0.0909)\,0.423\ \text{ohms} = 0.0385$$

$$Z_A = (0.0909)\,7.20\ \text{ohms} = 0.654$$

These values are shown in Figure 6.2. They are on a base of 4.16 kV.

Other requirements such as regulation, or surges on trips, do not enter in setting these values. As explained earlier, the worst-case trip is the generator tripping, and here the auxiliary bus is transferred to its start-up supply. Most of these loads then also shut down.

---

## Values on the Transformer's Nameplate

These values neglect resistive values. Including them will increase the previously calculated wye values slightly. For example, if the efficiency of the generator-to-high-voltage transformation were 98.7%, the losses are $(100\% - 98.7\%) = 1.3\%$. Then the 8% value increases to $[8.0^2 + 1.3^2]^{1/2} = 8.1\%$. This is within accuracy for the $Z_{H-G}$ measurement.

The nameplate values will reflect the standard test procedure methods. Measuring $Z_{H-G}$ is with a variable voltage source of 138 kV in the test lab, also used for other procedures on the transformer. Likewise, measuring $Z_{G-A}$ would be with a 13.8 kV variable source. Thus, nameplate values would be structured as follows:

| | |
|---|---|
| Hi-V winding to gen. winding | 8% of 20 MVA, on 138 kV base |
| Hi-V winding to aux. winding | 8% of 2 MVA, on 138 kV base |
| Gen. winding to aux. winding | 8% of 2 MVA, on 13.8 kV base |

$$I_{rated} = 18{,}000\ \text{kVA}/(1.732)(138\ \text{kV}) = 75.3\ \text{A}$$

$$Z = 0.08\,(Z_{rated}) = 0.08\big[(138\ \text{kV})/1.732\,(75.3\ \text{A})\big] = 0.08\,(1057\ \text{rated}) = 84.65\ \text{ohms}$$

This is the impedance in ohms of the transformer on the high-voltage side. The square of the turns ratio is 100, and it is 8% on a 138 kV base as shown.

These values may be specified for purchasing the transformer, but the manufacturer may bid slightly different values, or state wider tolerances than standard for his test results. These differences would result from other transformer design requirements, like BIL values, for example, and including resistive values.

For larger generating units, fault-current limits in the auxiliary distribution may require impedance values to be designed differently. The method used earlier was to base winding leakage reactive values on those of comparable transformers. Using the earlier equations, designing impedance values to meet fault-limiting requirements would be necessary.

## Fault-Current Analysis for the Protective Relay System

Until all fault currents have been evaluated, the transformer specification is incomplete. Looking at the contribution of the start-up transformer to a fault on the *4160 V bus*

$$I_{rated} = 2000\,kVA/(1.732)(4.16\,kV) = 278\,A$$

$$Z_{SU} = 0.08\big[(4160)/(1.732)\,278\big] = 0.691\,ohms\ on\ 4160\,V$$

$$Z_G = 0.0347$$

$$Z_H = 0.0385$$

$$Z_A = 0.654$$

The Thevenin impedance of the 138 kV system is calculated from the value in Figure 6.1, 5.9 kA on 138 kV: $Z_{TH} = (138\,kV/1.732)/5.9\,kA = 13.5\ ohms$.

$$Z_{TH} = (4.16\,kV/138\,kV)^2 / 5.9\,kA = 0.0123\ ohms\ on\ a\ 4.16\,kV\ base$$

$$Z_{TH} = 0.0123$$

$$Z_{GEN} = 0.147$$

The motor composite has an X" value of 16% on a 1000 kVA base. This converts to

$$I_M = 1000\,kVA/(4.16\,kV)(1.732) = 139\,A$$

$$Z_M = 0.16[(4.16\,kV)/(1.732)(139\,A)] = 0.16(17.27) = 2.76\ ohms\ on\ 4.16\,kV$$

These values have been entered in Figure 6.2.

The considered fault for the 4160 V diagram is a fault on the feeder supplying Z480, the 480 V transformer. (In Figure 6.2, move the fault cross up to the top of Z480.) This fault current is interrupted by the 600 ampere breaker on the 4160 V bus. By finding the paralleled value of $Z_{SU}$ with the motor composite, the value is 0.553 ohms. The fault contribution from this leg is then

$$I_{FAULT} = 2.4\,kV/0.553 = 4.39\,kA\ from\ this\ leg$$

Next, there are two series strings to evaluate, $Z_{GEN}$ in series with $Z_G(0.147 + 0.0347 = 0.182)$ and $Z_{TH}$ in series with $Z_H(0.0123 + 0.0385 = 0.0508)$.

By finding the paralleled value of these two series strings, the value is 0.0397 ohms. This paralleled value is then in series with $Z_A$. This series value is $(0.0397 + 0.654 = 0.694)$. The fault contribution from this leg is then

$$I_{FAULT} = 2.4\,kV/0.694 = 3.46\,kA\ from\ this\ leg$$

The two legs add to the fault current of 4.39 and 3.46 kA or a total of 7.85 kA.

This fault analysis shows that the 4160 V circuit breakers are stressed well below their interrupting ratings of 26 kA.

## Larger Units

As stated earlier, this fault current is low. It is compared to the interruption rating of the breakers on the 4160 V auxiliary bus. Here, the breaker rating is about three times the fault-current value. But for a larger unit (the largest unit is about 200 MVA), everything gets bigger and the impedances get smaller. The fault currents get much bigger and push interrupting ratings. Now, the transformer impedances must be adjusted to keep the fault currents within ratings. Also, the start-up power may come from another source, such as a local intermediate bus.

The impedances in this chapter are based on typically available transformers. The fault current through $Z_A$ comes mostly from the high-voltage source (compare $Z_{GEN}$ with $Z_{TH}$). Hence, $Z_{H-A}$ is likely to be increased. Balancing equations, with allowable fault currents, are written about the wye neutral. The neutral doesn't physically exist, and wye values can be negative. The analytical means, however, are given in the earlier equations.

# Section III

# Fast Transient Calculations

# 7

## Analysis of Power Cables Conducting Fast Transient Loads

Fast transient loads are kiloampere spikes that last several minutes. Rules for sizing power cables in buildings and plants are given in the *National Electric Code* (NEC). This code is adopted by almost all cities and towns in the United States and is followed. Rules for sizing power cables are based on the cables' loads.

The rules in the *National Electrical Code* assume that the loads are steady state. For varying loads, one must follow rules in the IEEE Buff Book, IEEE Std. 242 *Recommended Practice for Protection and Coordination of Industrial and Commercial Power Systems*. Here, only loads varying on a daily cycle are addressed. Presented in the following is a discussion of loads that are non-steady-state and that last for several minutes. A method is described for sizing power cables for these fast transient loads.

In many applications are found kiloampere loads that last for only a few minutes. A cable will heat up about 30°C in a few minutes. If the cable starts at about 60° and heats to 90°, the cable is properly sized for a fast transient load—that is, given that the insulation is rated for 90°.

After the kiloampere spike has happened, the cable must be allowed to cool back down to its starting temperature. The process that demands the kiloampere spikes dictates the current profile, and the current profile over several hours defines the conductor's sizing requirements.

For example, in a paper mill, the bark is milled off 4-ft-long logs, and chippers reduce the logs. The chips are sent to the digester. The debark mill is a large drum that is loaded with the logs. It rotates until the bark is sheared away by friction from the logs' tumbling. It takes the chipper about 3 s to reduce a 4 ft log. The chipper is driven by a motor, which sees a current spike for 3 s. But the next log is fed to the chipper in less than a minute, and the logs continue to arrive until the debark mill is emptied. The mill is then refilled with another batch of logs. It takes about 30 min for the mill to debark the new batch.

The chipper motor's current profile is spikes lasting 3 s. But there isn't enough time between spikes for a conductor to cool down. The motor's cable step-heats in 3 s steps until the debark mill is emptied. The spikes can be addressed as an rms value for chipping all the logs arriving out of the debark mill. The spikes are followed by an idle period of about 30 min. An overall worst-case rms value for the whole cycle, spikes and idle period, can then be defined. The motor, with its slow thermal response, sees only this rms current for the batch of logs. The motor's conductors see the spikes and the overall rms current value.

Another example is a combustion-turbine starting system in a combined cycle power plant, wherein the generator is used as a starting motor for the combustion turbine. The combustion turbine's exhaust is ducted to a heat-recovery boiler. A variable-frequency drive is employed that ramps up the shaft to a low speed and holds at this load, consisting of the engine's air compressor, for a purging cycle. Then the variable-frequency drive further ramps up the shaft speed to about 70%, when fuel is introduced and ignited. At this point, the turbine assumes the load and the electric motor current falls to zero. The duration of this starting cycle can be about half an hour for a large machine, and the motor current may reach 1600 A during the short period when fuel is introduced and ignited. This half-hour cycle is too short a period for conductors to reach their steady-state ratings. Even with consecutive start attempts, cables might not reach their steady-state ratings.*

---

## Cables in Air

If after a long time, a three-core power cable is energized, its temperature will climb to a steady-state value. Considering a small cable in free air without wind, this cable will heat up in about an hour. For cables in air, a single small bare cable in free air carrying a current will heat up exponentially with time:

$$\Phi = \Delta T\left(1 - e^{-t/K}\right) \qquad (7.1)$$

where
    $\Delta T$ is steady-state temperature rise on Kelvin or Celsius scales (°C)
    $\Phi$ is time-varying rise (°C)
    t is time (h)
    K is time constant equal to R times C (h)
    R is thermal resistance between wire and free air (°C/W)
    C is thermal capacitance of wire (W-h/°C)

---

* References in this book to the *National Electrical Code* are to the 2002 edition; table captions in Article 310 are less complicated in this edition.

This neglects any temperature difference within the cable's cross section and assumes that the cable is long enough that the temperature is uniform along the cable's length.

Here

$$\Delta T = Q R \tag{7.2}$$

where Q is constant heat loss in the cable (W).

If the current is constant, in (7.2), Q would be assumed constant by neglecting the rise in cable electrical resistance with temperature increase. But over a moderate value of temperature rise, this is a small change, and the average value of resistance over the value of $\Delta T$ may usually be used, or alternatively, the resistance at an adjusted temperature.

If the cable is insulated and if the small radial temperature difference through the insulation is neglected or taken as averaged, Equation 7.1 is true, but C must also include the insulation's thermal capacitance. Also, if the amount of cable is a unit length, then the values of R, C, and Q are per unit length. The effective value of $\Phi$ is said to be equal to $\Delta T$ when time t reaches a value of 4 K, or four time constants—that is, when the value of the exponential is 0.982. This is less than a 2% difference from the value of the exponential at infinite time.

If there are three conductors spaced together, with each carrying the same current as the conductor mentioned earlier, there is less than three times the exposed cable surface area in the bundle. Then the value of R for the bundle is greater than one-third the thermal resistance of one wire, so the bundle heats up to a slightly higher temperature rise than for the single wire. But the capacitance for the bundle is three times the thermal capacitance of one wire, so the value of K is slightly greater than for the single wire. Thus, the bundle heats up with a slightly longer heating time to the higher temperature rise. Also, this assumes that the small temperature differences through the cross section of the bundle are averaged. Given the rated temperature rise found in ampacity tables in the *National Electrical Code* [see NEC 2002, Table 310.67], the conductor current rating is found in these tables. The temperature rise with a lesser constant current value is proportional to the square of the current ratio if cable dielectric loss and shield loss are negligible. The applicable values for C can be found from manufacturers' data and the properties of the materials making up the cable.

An example of table values is, given a 500 kcmil 5 kV single unshielded copper conductor. A 1 ft length is considered. The ampacity for this conductor in air is 685 A (Table 310.60(C)(69)). The resistance is 0.027 ohms/1000 ft (NEC Table 9 Chapter 9). Based on these values, Q is 12.65 W (per ft) and R is 3.69°C/W (per ft). Based on the manufacturer's data sheets, the weight of copper is 1.54 lb and the weight of the insulation and jacket is 0.511 lb (both per ft). The specific heat of copper is 0.092, and the specific heat of polyethylene is 0.50. Converting units, the value of C for the copper

**TABLE 7.1**

K Factors in Table 43 for Equations in 8.5.2.4

|  | Air | | | |
| Cable | No Cond | In Cond | UG Duct | Direct Buried |
| --- | --- | --- | --- | --- |
| <#2 | 0.33 | 0.67 | 1.00 | 1.25 |
| #2–4/0 | 1.00 | 1.50 | 2.50 | 3.0 |
| ≥250 MCM | 1.50 | 2.50 | 4.00 | 6.00 |

*Source:* IEEE STD 242-86. Copyright 2002 IEEE. With permission.

is 0.0749 W-h/°C and the value of C for the insulation is 0.1347 W-h/°C. Their sum *C is 0.2096 W-h/°C.* (Note that the copper is about one-third of the total.) Multiplying these values for R and C for this large single conductor in free air, the value for K is 0.77 h. It would take a period of about four time constants, or 3.1 h, for this conductor to heat up.

Considering a 1 ft length of three bundled conductors of this same cable in free air, the ampacity listed in Table 310.60(C) (67) is 645 A. Q is 33.7 W, the table temperature rise is 65C, and *R is 1.93°C/W.* The value of C is three times the value of one conductor: *C is 0.629 W-h/°C.* With these values for R and C for these three large single-bundled conductors in free air, the value for K is 1.21 h. It would take a period of about four time constants, or about 5 h, for this three-conductor bundle to heat up.

Empirical values for K are listed in Table 43 of ANSI/IEEE Std. 242-86. Table 43 applies to three-conductor cables and is repeated here in Table 7.1.

The application for these table values in this standard is similar to transient temperature calculations. The table values are suitable for the application in the standard, in section 8.5.2.4, which makes use of the values for K. The value of 1.21 h for the said 500 kcmil cables compares with the table value of 1.50 h.

Note in this table that comparing cables in air and cables in conduit in air, the latter have a longer K factor, some one and a half times or so longer.

## Cables Installed in Conduit

If the bundle is installed in a galvanized conduit in free air, (7.1) becomes more complicated. Now, there is a resistance between the bundle and the conduit, and a second resistance between the conduit and free air. Also, the conduit's thermal capacitance affects the heating time. First, because of proximity effect, the resistance has increased, increasing the heating time. Second, the conduit doesn't begin to heat up in the beginning. It doesn't have heat generated in it like the bundle, so it begins to heat up after the bundle starts to transfer heat to it, which happens only as the bundle heats up. And the conduit's thermal capacitance adds to that of the system. With both resistance and system capacitance increased, the time constant is increased. (Here again, the small radial temperature difference through the conduit wall is neglected.)

In Appendix A is derived the exponential function for the temperature rise (A.18c). There are now two time constants, $K'$ and $K''$, with $K'$ greater than $K''$. The formulas for the time constants are (A.13) through (A.15) in Appendix A.

In Appendix A, the thermal resistance between the bundle and conduit is R1. The thermal capacitance of the conductors is $C_1$. The thermal resistance between the conduit and its ambient is $R_2$, and the thermal capacitance of the conduit is $C_2$.

From (A.18c), a simplification may be developed based on the criterion Ch. The following is the definition of the criterion Ch.

It can be shown that $\dfrac{K_2 R_1}{R_1 + R_2} = K'' \dfrac{K'}{(R_1 + R_2) C_1}$. Then

$$Ch \equiv \frac{K'}{(R_1 + R_2) C_1} \tag{7.3}$$

The significance of Ch is that it is the ratio of the actual system time constant, $K'$, to *the apparent time constant*, $(R_1 + R_2)C_1$. If there were no conduit, as in the 500 kcmil example earlier, there would be no $R_2$, and the time constant would be $R_1 C_1$. The denominator of Ch is the apparent time constant with the conduit. The system time constant, $K'$, is always greater than the apparent time constant. The apparent time constant is used when only the sum, $R_1 + R_2$, is known, that is, when either $R_1$ or $R_2$ cannot be found. (See the motor thermal model in Chapter 11.)

The sum, $R_1 + R_2$, is the thermal resistance of the conductors to ambient air. This sum is the temperature rise of the conductors divided by the watts generated in the three conductors. Both the temperature rise and ampacity can be determined from Table 310.60(C)(73) in the NEC for the three conductors installed in a conduit. The watts generated in 1 ft of the three conductors is three times the current squared in each conductor times each conductor's electrical resistance. (The electrical resistance is listed in NEC Table 9 of Chapter 9.) The listed values include proximity effect. (Dielectric and shield losses must be negligible.) Then the temperature rise listed in the table is divided by the watts, yielding $R_1 + R_2$. This quotient is, in turn, multiplied by the thermal capacitance of the three cables, $C_1$, for the apparent time constant.

The thermal capacitance is the capacitance of 1 ft of the three-conductor cable. The capacitance $C_1$ is three times the sum of 1 ft of the copper's thermal capacitance plus its insulation's and jacket's thermal capacitance. Thermal capacitance is weight times specific heat.

Table values of the time constants are developed in Table 7.3. Included are values for the criterion Ch. Based on these values, the following definition and development are valid. The values for Ch in Table 7.3 increase with smaller conductor size. For these power cable sizes, Ch values are all less than two.

Let $1 < Ch < 2$, then $Ch \equiv 1 + a$ fraction, or
$Ch \equiv 1 + Fh$, where Fh is the fractional part of Ch. Then

$$\frac{K_2 R_1}{R_1 + R_2} = K''(1 + Fh) = K'' + K''Fh$$

$$\text{In } \Phi_c = \Delta T \left[ 1 - \frac{1}{K' - K''}\left( K' - \frac{K_2 R_1}{R_1 + R_2} \right) \right] e^{-\frac{t}{K'}} + \Delta T \left[ \frac{1}{K' - K''}\left( K'' - \frac{K_2 R_1}{R_1 + R_2} \right) \right] e^{-\frac{t}{K''}}$$

(A.18c)

$$\Phi_c / \Delta T = 1 - \frac{1}{K' - K''}\left[ K' - \left( K'' + K''Fh \right) \right] e^{-\frac{t}{K'}} + \left[ \frac{K'' - K''}{K' - K''} + \frac{-K'Fh}{K' - K''} \right] e^{-\frac{t}{K''}}$$

$$\Phi_c = \Delta T \left( 1 - e^{-\frac{t}{K'}} \right) + \Delta T \frac{K''Fh}{K' - K''}\left( e^{-\frac{t}{K'}} - e^{-\frac{t}{K''}} \right)$$

(7.4)

This is an expression of the simple exponential plus an adjustment, calculated from Fh, the fractional part of Ch. If you look closely at the adjustment, you will find a hump curve. It has an amplitude and two exponentials. Looking at the two exponentials, at $t$ equal to zero, each is equal to one. Thus, their difference starts at zero, grows to a maximum after about two K"s, and decays. Fast transients happen during early intervals of (7.4) where the simple exponential and the correction terms are added.

The correction is a significant added adjustment during early intervals.

## Underground Feeders

Burying the conduit containing the bundle of three wires increases the system thermal resistance, adding now, instead of the resistance between the conduit and free air, the increased resistance between the conduit and through the path of heat flow to the earth's surface. Also, the thermal capacitance of the earth in this path increases the overall amount of capacitance. These additions of thermal resistance and capacitance increase yet again the system time constant and further complicate the system exponential response. If the added complication from installing the cables in a conduit in free air is projected to burying the conduit, then more coefficients and more time constants in the expression of the time-varying temperature rise could be expected, as listed in Table 7.1.

The *National Electrical Code* (NEC) table ampacities are based on published electrical resistance values. The code states that adjustments must be made for shield and dielectric losses. Also, adjustments must be made for heating from harmonic currents.

## Calculation of Time Constants

The calculation tables, Tables 7.3 and 7.4, illustrate the equations of Appendix A. In these calculations, Q, the heat flow (in watts) generated in the three conductors, is based on the conductor electrical resistance listed in the NEC. Thermal capacitance, C in the earlier text, is $C_1$ in the two calculation tables, because $C_1$ represents the thermal capacitance of the conductors and their jackets whether in air, in tray, or in conduit. Thus, the time constant RC is $R_t C_1$ in Table 7.3.

The electrical resistance, Re in the table, is taken from the NEC, Chapter 9 Table 9. For bundled cables in tray or free air, the thermal resistance R in the earlier text is labeled Rt in the table. For three single cables in conduit, the value for $R_1$ is obtained by calculating $R_1 + R_2$ by using the ampacity table cited in the calculation sheet and subtracting $R_2$.

The value of $R_2$ is found by multiplying Rt by the ratio of the triplexed conductors' diameter divided by the conduit diameter. The diameter of a circle that circumscribes the three conductors is used to represent the triplexed conductors' diameter. This circle's diameter is 2.155 times one conductor's diameter. No adjustments were made for emissivity, or diameter in the conduit's convective coefficient, comprising $R_2$, because these were considered as almost the same as for the circumscribing circle. The value for $R_1$ determined in this manner is almost always less than $R_t$. This implies that

**TABLE 7.2**

Summary Cable Heating Tabulation

| | Calculation of Temperature Rises | | | |
|---|---|---|---|---|
| | 5 kV Unshielded Polyethylene Insulation and Jacket | | | |
| | 250 kcmil | 350 kcmil | 500 kcmil | 750 kcmil |
| | In open tray or free air | | | |
| 1000 A for 30 s | 2.96°C | 1.69°C | 1.07°C | 0.57°C |
| 1000 A for 5 min | 28.5°C | 16.4°C | 10.4°C | 5.58°C |
| 2000 A for 30 s | 11.8°C | 6.76°C | 4.3°C | 2.29°C |
| 2000 A for 2.5 min | 58.2°C | 33.3°C | 21.2°C | 11.3°C |
| 2000 A for 5 min | 114.2°C | 65.5°C | 41.6°C | 22.3°C |
| 2000 A for 10 min | 219.7°C | 126.5°C | 80.5°C | 43.5°C |
| | In steel conduit in free air | | | |
| 1000 A for 30 s | 3.1°C | 1.74°C | 1.15°C | 0.58°C |
| 1000 A for 5 min | 29.4°C | 16.6°C | 11.0°C | 5.5°C |
| 2000 A for 30 s | 12.3°C | 6.67°C | 4.6°C | 2.33°C |
| 2000 A for 2.5 min | 60.3°C | 34.0°C | 22.5°C | 11.2°C |
| 2000 A for 5 min | 117.5°C | 66.6°C | 44.1°C | 22.2°C |
| 2000 A for 10 min | 223.9°C | 127.8°C | 84.7°C | 43.0°C |

*Source:*  Henry, R.E., Response of power cables to fast transient loads, in *IEEE Ind. Appl. Soc.*, Seattle, WA. Copyright 2004 IEEE. With permission.

the snug fit of the cables in the conduit results in an air gap that is based on the cited ampacity table.

Table 7.2 is a summary of calculation sheets, of which Table 7.4 is an example. The temperature rises, calculated from the values in the results of Table 7.3 and as given in Table 7.4 and other like tables, have been summarized in Table 7.2.

The conduit size is determined as for 40% fill or the next larger size. Thus, a 3½-size Schedule 40 conduit is used for 250 kcmil, a 4 size for 500 kcmil, and a 5 size is used for 750 kcmil. In the ampacity tables, no adjustment in ampacity from using diameters larger than these sizes is made.

The values for specific heat in the table are converted from water, having a specific heat of 0.5274 W-h/lb °C. The relative specific heat of polyethylene, of 0.50, was used in the tables for insulation and jacket. This value is for a polymer temperature of 75°C. At 20°C, sources list a value of 0.55 for polyethylene.

In Tables 7.3 and 7.4, the values for K and K′ are calculated as discussed earlier. Comparing the values with those listed in Table 43 of ANSI/IEEE Std. 242-86 (shown in Table 7.1), the values in Table 43 are a fair summary of the values in Tables 7.3 and 7.4.

The values of K in Table 43 validate the values in Tables 7.3 and 7.4.

**TABLE 7.3**

Cable Heating Tabulation

| | | Calculation of Constants | | | |
|---|---|---|---|---|---|
| Constant | Source | 250 MCM | 350 MCM | 500 MCM | 750 MCM |
| | | 5 kV | 5 kV | 5 kV | 5 kV |
| | | In open tray or free air | | | |
| Re ohms/1000′ | Code: Ch 9 Tbl 9 | 0.052/1000′ | 0.038/1000′ | 0.027/1000′ | 0.023/1000′ |
| I3 (105°C) | Code: Tbl 310-67 | 415 A | 515 A | 645 A | 835 A |
| 3 I3sqrd Re = Q3 | | 27.90 W | 31.03 W | 36.19 W | 43.93 W |
| Temp/Q3 = Rt | Code: Tbl 310-67 | 2.33°C/W | 2.09°C/W | 1.80°C/W | 1.48°C/W |
| Wconductor lb | Data sht. | 3.555 lb | 4.77 lb | 6.165 lb | 9.060 lb |
| Wcu lb | Data sht. | 2.316 lb | 3.243 lb | 4.632 lb | 6.848 lb |
| ccu | Data sht. | 0.0485 | 0.0485 | 0.0485 | 0.0485 |
| Wcu ccu W-h/°C | | 0.1724 | 0.1573 | 0.2247 | 0.3321 |
| Wcond - Wcu = Wi lb | | 1.239 lb | 1.527 lb | 1.533 lb | 2.212 lb |
| ci W-h/lb °C | Data sht. | 0.264 | 0.264 | 0.264 | 0.264 |
| Wi ci W-h/°C | | 0.3531 | 0.4031 | 0.4047 | 0.5839 |
| Wccu + Wici = $C_1$ | | 0.436 W-h/°C | 0.5604 W-h/°C | 0.629 W-h/°C | 0.9161 W-h/°C |
| $R_tC_1$ | | 1.02 h | 1.17 h | 1.13 h | 1.36 h |
| | | In steel conduit in free air | | | |
| Conductor Dia. Dc | Data sht. | 1.14 in. | 1.24 in. | 1.40 in. | 1.64 in. |
| 2.155 Dc = De | | 2.457 in. | 2.672 in. | 3.017 in. | 3.534 in. |
| 2.74 Dc = Dd min | | 3.12 in. | 3.40 in. | 3.84 in. | 4.49 in. |
| Conduit Dia. ODd | Data sht. | 4.000 in. | 4.000 in. | 4.500 in. | 5.563 in. |
| De/ODd Rt = $R_2$ | | 1.43°C/W | 1.40°C/W | 1.21°C/W | 0.940°C/W |
| Re ohms/1000′ | Code: Ch9 Tbl 9 | 0.054/1000′ | 0.039/1000′ | 0.029/1000′ | 0.021/1000′ |
| I3 (105°C) | Code: Tbl 310-67 | 415 A | 515 A | 645 A | 835 A |
| 3 I3sqrd Re = Q3 | | 27.90 W | 31.03 W | 36.19 W | 43.93 W |
| Id (105°C) | Code: Tbl 310-73 | 355 A | 430 A | 530 A | 665 A |
| 3 Id sqrd Re = Qd | | 20.416 W | 21.633 W | 24.44 W | 27.86 W |
| Temp/Qo= $R_1 + R_2$ | Code: Tbl 310-73 | 3.18°/W | 3.005°/W | 2.66°/W | 2.33°/W |
| $R_1$ | | 1.75°C/W | 1.605°C/W | 1.45°C/W | 1.39°C/W |
| Conduit Wt. Wd | Data sht. | 8.31 lb | 8.31 lb | 9.82 lb | 13.44 lb |
| cd | Data sht. | 0.0567 | 0.0567 | 0.0567 | 0.0567 |

*(Continued)*

**TABLE 7.3 (*Continued*)**

Cable Heating Tabulation

| Constant | Source | Calculation of Constants | | | |
|---|---|---|---|---|---|
| | | 250 MCM | 350 MCM | 500 MCM | 750 MCM |
| | | 5 kV | 5 kV | 5 kV | 5 kV |
| Wd cd = $C_2$ | | 0.471W-h/°C | 0.471W-h/°C | 0.557W-h/°C | 0.762 W-h/°C |
| $R_1 C_1 = K_1$ | | 0.763 | 0.899 | 0.912 | 1.273 |
| $R_2 C_2 = K_2$ | | 0.674 | 0.659 | 0.674 | 0.716 |
| $R_2 C_1 = K_3$ | | 0.623 | 0.785 | 0.761 | 0.861 |
| $2K_1 K_2$ | | 1.0285 | 1.185 | 1.229 | 1.823 |
| $K_1 + K_2 + K_3$ | | 2.060 | 2.343 | 2.347 | 2.850 |
| $(K_1 + K_2 + K_3)^2$ | | 4.2436 | 5.4896 | 5.508 | 8.123 |
| $4K_1 K_2$ | | 2.0570 | 2.37 | 2.458 | 3.646 |
| Rho | | 1.4787 | 1.7662 | 1.747 | 2.116 |
| $K'$ | | 1.77 h | 2.05 h | 2.05 h | 2.48 h |
| $K''$ | | 0.291 h | 0.288 h | 0.30 h | 0.37 h |
| $K'/(K_1 + K_3) = Ch$ | | 1.276 | 1.217 | 1.225 | 1.162 |
| $4K'$ | | 7.1 h | 8.2 h | 8.2 h | 9.9 h |

*Source:* Henry, R.E., Response of power cables to fast transient loads, in *IEEE Ind. Appl. Soc.*, Seattle, WA. Copyright 2004 IEEE. With permission.

## Cooldown

The formulas in Appendix I describe warm-up. Cooldown is similar. If the simple exponential function for increasing response is $\Delta T (1 - e^{-t/K})$, then the simple exponential function for the decay response is $\Delta T e^{-t/K}$. However, there will be two time constants in the formulas for the cooldown exponential, $K'$ and $K''$, which will have the same values as in Table 7.4. The *Ch* criterion will still be valid, and just as $\Phi c$ reduces to the simple exponential, the cooldown expression would also reduce to the approximation $\Phi c = \Delta T e^{-t/K'}$.

To the extent that Ch does not approach one, the decay can be expressed as the sum of two exponentials, with time constants of $K'$ and $K''$, and with coefficients that add to the initial cable temperature. The two exponential expressions can be illustrated by constructing a semilog plot of the cable cooldown from initial temperature rise, on the log-scale ordinate. Starting from initial time of the cooldown, the plot is the sum of two lines, each an exponential decay, with slopes of $-K'$ and $-K''$. The line with slope $-K'$ is similar to $\Phi c$ mentioned earlier, except the sum of the initial temperatures of both lines equals $\Delta T$. Here, $\Delta T$ is the rise reached at the end of the warm-up. Just as the warm-up could take hours for its response, so does the cooldown.

Cooldown has the same time constant as warm-up.

**TABLE 7.4**

Cable Heating Tabulation

| | Calculation of Spikes | | | |
|---|---|---|---|---|
| **Constant** | **250 MCM** | **350 MCM** | **500 MCM** | **750 MCM** |
| | **5 kV** | **5 kV** | **5 kV** | **5 kV** |
| | *2000 A for 30 s* | | | |
| | In open tray or free air | | | |
| Re ohms/1000' | 0.052/1000' | 0.038/1000' | 0.027/1000' | 0.021/1000' |
| W/ft | 624 W | 456 | 324 | 252 |
| Rt | 2.42°/W | 2.09°/W | 1.93°/W | 1.64°/W |
| Delta T | 1510°C | 953.0°C | 625.3°C | 413.3°C |
| K | 1.06 h | 1.17 h | 1.21 h | 1.50 h |
| $1 - \exp -t/K$ | 0.0078308 | 0.0070972 | 0.0068634 | 0.0055402 |
| Peak °C | 11.8°C | 6.76°C | 4.3°C | 2.29°C |
| | In steel conduit in free air | | | |
| $R_1 + R_2$ | 3.18°/W | 3.005°/W | 2.66°/W | 2.33°/W |
| Delta T | 2060.8° | 1406.4° | 925.6° | 587.2° |
| *K'* | 1.77 h | 2.05 h | 2.05 h | 2.48 h |
| *K''* | 0.291 h | 0.288 h | 0.30 h | 0.37 h |
| Rho | 1.4787 | 1.7662 | 1.747 | 2.116 |
| Fh | 0.276 | 0.217 | 0.225 | 0.162 |
| K''Fh/Rho | 0.05432 | 0.035384 | 0.03864 | 0.02775 |
| $1 - \exp -t/K'$ | 0.004697 | 0.004568 | 0.0040568 | 0.0033546 |
| $\exp -t/K''$ | 0.97177 | 0.97148 | 0.9726 | 0.9777292 |
| $\exp -t/K' - \exp -t//K''$ | 0.02353 | 0.02395 | 0.02334 | 0.02227 |
| exp + adj | 0.005975 | 0.004746 | 0.004959 | 0.003973 |
| Peak °C | 12.3°C | 6.67°C | 4.6°C | 2.33°C |
| | *2000 A for 2.5 min* | | | |
| | In open tray or free air | | | |
| Re ohms/1000' | 0.052/1000' | 0.038/1000' | 0.027/1000' | 0.021/1000' |
| W/ft | 624 | 456 | 324 | 252 |
| Rt | 2.42°/W | 2.09°/W | 1.93°/W | 1.64°/W |
| Delta T | 510° | 953.0°C | 625.3°C | 413.3°C |
| K | 1.06 h | 1.17 h | 1.21 h | 1.50 h |
| $1 - \exp -t/K$ | 0.03855 | 0.03499 | 0.03385 | 0.027395 |
| Peak °C | 58.2°C | 33.3°C | 21.2°C | 11.3°C |
| | In steel conduit in free air | | | |
| $R_1 + R_2$ | 3.18°/W | 3.005°/W | 2.66°/W | 2.33°/W |
| Delta T | 2060.8° | 1406.4° | 925.6° | 587.2° |
| *K'* | 1.77 h | 2.05 h | 2.05 h | 2.48 h |
| *K''* | 0.291 h | 0.288 h | 0.30 h | 0.37 h |

*(Continued)*

**TABLE 7.4 (*Continued*)**

Cable Heating Tabulation

| | Calculation of Spikes | | | |
|---|---|---|---|---|
| **Constant** | **250 MCM** | **350 MCM** | **500 MCM** | **750 MCM** |
| | **5 kV** | **5 kV** | **5 kV** | **5 kV** |
| Rho | 1.4787 | 1.7662 | 1.747 | 2.116 |
| Fh | 0.276 | 0.217 | 0.225 | 0.162 |
| K"Fh/Rho | 0.05432 | 0.035384 | 0.03864 | 0.02775 |
| $1 - \exp -t/K'$ | 0.02327 | 0.02012 | 0.02012 | 0.016661 |
| $\exp -t/K''$ | 0.86659 | 0.8653 | 0.87032 | 0.8935 |
| $\exp -t/K' - \exp -t//K''$ | 0.11014 | 0.11458 | 0.10956 | 0.089839 |
| $\exp + adj$ | 0.029253 | 0.024174 | 0.024353 | 0019154 |
| Peak °C | 60.3°C | 34.0°C | 22.5°C | 11.2°C |

*Source:*  Henry, R.E., Response of power cables to fast transient loads, in *IEEE Ind. Appl. Soc.*, Seattle, WA. Copyright 2004 IEEE. With permission.

## Simple Fast Transient Response

Table 7.2 lists the temperature rises in four power cables for the fast transient spikes listed in the left-hand column. It summarizes the calculations in Tables 7.3 and summarizes the calculations in Table 7.4 and five other tables that Table 7.4 illustrates.

Several results of the calculations shown in Table 7.2 are notable. First, the temperature rise is *proportional* to time for each short spike in a given cable. Until the duration of the spike exceeds 5 min in the 250 and 350 kcmil sizes, the rises with either current value are proportional to the duration. This can be expected because the early part of the exponential curve is essentially linear in its beginning portion. Five minutes represents only one-twelfth of the first time constant (K for three 250 kcmil conductors in air is 1.02 h), which is very early in the response of a 250 kcmil feeder in air or spaced in tray.

The second result is that the rise appears to be the same whether the bundle is in free air or spaced in tray or in conduit in air. The values for the bundle in free air differ from the values in conduit by a small amount, about 5%. This can be expected because the generated heat is slightly different when in conduit, owing to a different proximity effect, and the values are calculated by different formulas. Only in the values for a 10 min duration do differences appear that exceed 5%. For the 750 kcmil cable size, with its longer time constant, the difference does not appear in the 10 min duration. This implies that a simple relationship, rather than the formulas used to calculate the rises, exists for these short heavy spikes.

A fast transient load's kiloampere current spike only lasts several minutes. Then the transient temperature rise happens in the early portion of its exponential response.

- The region of fast transient response is within the first few minutes of the first time constant of a three-conductor power cable.
- Five minutes is one-thirteenth of the first time constant of a 250 kcmil feeder in air.
- Ten minutes is one-sixth. Ten minutes is one-sixteenth of the first time-constant response of a 750 kcmil cable in air.
- This early portion of the exponential response is essentially linear. The temperature rise from a fast transient load is therefore linear with time.

Heating by the current spike is assumed to begin at ambient temperature. The heating therefore begins at a temperature rise, $\Phi_c$, of zero. Then $\Phi_c$ is proportional to time. The slope of $\Phi_c$ is found by taking the derivative of $\Phi_c$ and setting time to zero.

The function of $\Phi_c$ in *open tray or free air* is, with $\Phi_c$ the temperature rise of the three cables above ambient temperature

$$\Phi_c = \Delta T\left(1 - e^{-t/K}\right)$$

where $\Delta T = Q\,R_t$

Q is the heat generated in 1 ft of the three cables by the kiloampere spike
$R_t$ is the thermal resistance of 1 ft of the three cables to the ambient
t is time in hours
K is $R_t\,C_1$, hours
$C_1$ is the thermal capacitance of 1 ft of the three cables, W-h/°C

$$\frac{d\Phi_c}{dt} = -\Delta T\left(-\frac{1}{K}\right)e^{-\frac{t}{K}}, \quad \text{and at } t = 0$$

$$\frac{d\Phi_c}{dt} = \Delta T/K = \left(\frac{QR_t}{R_tC_1}\right) = \frac{Q}{C_1}$$

and

$$\Phi_c = \frac{Q}{C_1}t \quad 0 \le t \ge 0.2\,K \text{ for the spike} \tag{7.5}$$

The function of $\Phi_c$ *in conduit in free air* is, with $\Phi_c$ again the temperature rise of the three cables above ambient temperature, Equation A.18a in Appendix A:

$$\Phi_c = \Delta T\left[1-\frac{1}{K'-K''}\left(K'e^{-\frac{t}{K'}}-K''e^{-\frac{t}{K''}}\right)\right]+R_1Q_0\frac{K_2}{K'-K''}\left(e^{-\frac{t}{K'}}-e^{-\frac{t}{K''}}\right) \quad \text{(A.18a)}$$

The derivative of (A.18a)

$$\frac{d\Phi_c}{dt}=-\frac{\Delta T}{K'-K''}\left(\frac{-K'}{K'}e^{-\frac{t}{K'}}-\frac{-K''}{K''}e^{-\frac{t}{K''}}\right)+R_1Q\frac{K_2}{K'-K''}\left(\frac{-1}{K'}e^{-\frac{t}{K'}}-\frac{-1}{K''}e^{-\frac{t}{K''}}\right)$$

At time $t = 0$

$$\frac{d\Phi_c}{dt}=-\frac{\Delta T}{K'-K''}\left(\frac{-K'}{K'}-\frac{-K''}{K''}\right)+R_1Q\frac{K_2}{K'-K''}\left(\frac{1}{K''}-\frac{1}{K'}\right)$$

$$=0+R_1Q\frac{K_2}{K'-K''}\left(\frac{1}{K''}-\frac{1}{K'}\right)$$

Using the relationship in Appendix A

$$\frac{1}{K''}-\frac{1}{K'}=\frac{K'-K''}{K_1K_2}$$

$$\frac{d\Phi_c}{dt}=R_1Q\frac{K_2}{K_1K_2}=R_1Q\frac{1}{K_1}=R_1Q\frac{1}{R_1C_1}=\frac{Q}{C_1}$$

which is the same as in open tray or free air.

- The rates of rise are identical, whether in free air, spaced in tray, or in conduit in air.
- The temperature rise from the current spikes depends on the rate of heat generated in the three cables and the cables' thermal capacitance.
- The rate of rise is independent of the conductors' thermal resistance to free air.
- Therefore, the temperature rise of conductors due to fast transient current spikes is the same for any installation of conductors, in air, in conduit in air, underground or direct buried.
- Equation 7.5 applies to all installations:

$$\Phi_c=\frac{Q}{C_1}t \quad 0\le t\ge 0.2\ K' \quad \text{for the spike} \quad \text{(7.5)}$$

## Power Cables Rated 600 V

Table 7.5 is for 600 V power cables. It is a summary table like Table 7.2. Tables have been created herein for responses of 5 kV power cables to spikes of kiloampere currents lasting up to 10 min. Tables 7.3 and 7.4, with other tables like Table 7.4, develop responses of 5 kV cables. Table 7.2 summarizes the values in Table 7.4 and other like tables. In creating Table 7.5, calculations using Equation 7.5 were employed. Data for the tables were drawn from manufacturers' literature and the NEC tables cited earlier.

Table 7.5 also includes a comparison of time constants for 5 kV and 600 V cables.

## Sizing Power Feeders Using the Tables

The value of the summary tables, Tables 7.2 and 7.5, is to rapidly compare cable temperature responses to a transient current profile and select a cable size. The method is developed in Chapter 8. Conductor sizing uses two elements, the rms equivalent current and the current spike's amplitude and duration.

**TABLE 7.5**

480 V Cable Heating Tabulation

|  | 250 kcmil | 350 kcmil | 500 kcmil | 750 kcmil |
|---|---|---|---|---|
| 1000 A for 30 s | 4.56°C | 2.70°C | 1.45°C | 0.80°C |
| 1000 A for 5 min | 45.6°C | 27.0°C | 14.5°C | 7.99°C |
| 2000 A for 30 s | 18.2°C | 10.8°C | 5.80°C | 3.12°C |
| 2000 A for 2.5 min | 91.2°C | 54.0°C | 29.0°C | 15.6°C |
| 2000 A for 5 min | 182.4°C | 108.0°C | 58.0°C | 31.2°C |
|  | VALUES for C1 (W-h/ft-°C) |  |  |  |
| 600 V EPR with CSPE jacket | 0.285 | 0.352 | 0.467 | 0.657 |
| 5 kV Unshielded XLP and jacket | 0.436 | 0.560 | 0.629 | 0.916 |

*Source:* Henry, R.E., Fast transient loads of low-voltage power cables, *Industrial and Commercial Power Systems 51st Tech Conference IEEE Industrial Applied Society,* Calgary, Alberta, Canada. Copyright 2015 IEEE. With permission.

*Notes:* 600 V EPR insulation and CSPE jacket; 65 mil insulation, 65 mil jacket; 80 mil and 65 mil for 750 kcmil.

## Cable Assemblies

Ampacity tables all represent that the bundled three conductors and their insulations are all at the same temperature, or the thermal gradients in the bundle are very small and they can be averaged. The definition of $C_1$ is also based on this representation.

The said model with conductors in conduit shows that between the bundled conductors and the conduit, a thermal resistance exists, R1. The NEC ampacity tables also show that cables in conduit have less ampacity than in free air (comparing Table 310.73 with Table 310.67); that is, with the addition of $R_2$, the cables' same temperature rise in the two tables takes less heat generated in the cables. Doing this same comparison between the cables in air and the cables in manufactured assemblies in air (comparing Table 310.67 with Table 310.71), the same reduced ampacity results with the assembly. And further, comparing Tables 310.73 and 310.75 (three singles in conduit and three conductors in a cable in conduit), less ampacity is seen with the cable assembly. This comparison shows that cable assemblies have more thermal resistance than a bundle of three conductors in free air and that the fillers, jackets, and sheaths of the assembly increase $C_1$.

In a cable assembly, the fillers, jackets, and sheaths add thermal resistance, which decouples these materials from the three conductors. Although the inner fillers in contact with the conductors would tend to add thermal capacitance and increase $C_1$, it is conservative to assume that no capacitance is added by the fillers. If so, the values of spikes of temperature rise in Table 7.2 represent conductors in cable assemblies. However, cooldown will take longer.

This analysis has developed the time constants of cables in air and in conduit in air. The values of the time constants are validated by Table 7.1 in the "Cables in Air" section. This analysis has developed a formula for temperature rise of cables conducting kiloampere currents for several minutes. It is now possible to use these formulas to calculate the temperature rise of power feeders subjected to fast transient loads. The next chapter presents methods to size power feeders for fast transient loads.

# 8

## Sizing Power Cables for Fast Transient Loads

Fast transient loads are kiloampere current spikes lasting several minutes. They result from an occasional high torque demanded by process machinery or a spurt of heating in a process that otherwise has a more moderate power profile. Ore crushers, log chippers, and turbine starting are examples. These are discussed in Chapter 7. This discussion assumes that the current profile can be defined as a worst-case requirement and that the included current spikes are occasional in the profile. A three-conductor power cable is required to conduct the three-phase current profile.

The method begins with developing the current–time profile required for the conductors. Typically, this profile might last about 12 h. During this period, occasional fast transient loads occur. Next, the rms value of the current profile is calculated. The current's rms value is equivalent to the same steady-state current. The large current spikes are then separated out, and the largest spike is assumed to occur at the profile's beginning. If there are more than one representative spike, a typical maximum spike can also be used.

Conductor sizing uses the two elements, the rms equivalent current and the biggest current spike. With a trial conductor size, the steady-state conductor temperature for this rms equivalent current is found from interpolating the NEC ampacity tables. The temperature response due to the spike is found by entering the summary, Table 7.2, or for 600 V cables, Table 7.5. The conductor temperature rise, resulting from the rms current, and the conductor temperature rise, resulting from the spike, are summed and compared with the conductor's temperature rating. The comparison shows whether the trial conductor size is adequate, or if a new trial size is indicated.

For example, if the rms current were 71% of the trial conductor's rated ampacity, the rms current's temperature rise would be half of the trial conductor's rated rise. That is, if its rated rise were 90°C in a 40°C ambient (a rise of 50°C), with a current of 71% of its ampacity, the conductor would have a rise of 25°C, because 0.71 squared is half. Then if the spike's interpolated temperature rise were also half of the trial conductor's rated rise, their sum would equal the conductor's rated rise, and the conductor would be a suitable size for the profile.

## Developing the Summary Tables 7.2 and 7.5

Table 7.2 summarizes the temperature rises of spikes in three-conductor 5 kV unshielded power cables of 250, 350, 500, and 750 kcmil. Each cable is subjected to spikes of 1000 and 2000 A for times of 30 s through 10 min. For cables in free air and in conduit in air, both are listed. The values can be interpolated in kiloamperes and in durations.

The choice of unshielded 5 kV cables was mainly because there are no shield losses with unshielded cables. Shield losses, if known, can be added to the kiloampere's watt values. Also, harmonic losses can be added. With a six-pulse harmonic profile and with increasing cable resistance with skin effect, the harmonic loss is 14.5% of the fundamental's watt loss in the cable. Thus, with the six-pulse harmonic content, the corrected kiloampere value for entering the table would be 107% of the fundamental current spike. (1.07 squared is 1.145, or 14.5% increase of the watt dissipation.)

If one compares the manufacturer's data for 5 kV cables, one finds that EPR is very slightly heavier than XLP, and jacket materials also slightly differ in weight. Also, the specific heat values vary slightly for these materials. Rather than refining values, conservative sizing is a better practice.

## Low Voltage

Using Equation 7.5 derived in Chapter 7, the time constants for 480 V cables were calculated and are listed in Table 7.5. Also listed for comparison are the time constants for the 5 kV cables listed in Table 7.3. Because the 480 V cables have thinner insulation and a thinner jacket than the 5 kV cables, their C1 values of thermal capacitances are less. And because the 480 V cables are thinner, they have less surface area and, hence, more thermal resistance. With lesser capacitance and more resistance, their product, K', is slightly less than the values for the 5 kV cables. (A listing for comparison of the capacitance values is given in Table 7.5.)

It is shown in Table 7.3 that the capacitance of a cable (C1) is made up more from the insulation and jacket than from the copper conductor. Although the copper may weigh more, its specific heat is less than that of the insulation and jacket. For an application of cable to be used for fast transient heating in Table 7.5, a large-capacitance component was chosen, that is, one having an ethylene propylene insulation and jacket. A large value for C1 means less temperature rise in the spike.

From these values, for the two smallest sizes of 250 and 350 kcmil, the one-seventh duration limit is 12 and 13 min. Then spikes of 10 min duration may be calculated by Equation 7.5.

## Fast Transient Loading of the Motor

Machinery discussed herein would have intervals of nominal torque requirements followed by high torque requirements for several minutes. Starting a combustion turbine is one example. Another example might be a gear crusher reducing an ore to smaller mesh sizes. A fast transient torque loading can require a lower shaft speed while crushing the ore. The motor driving the crusher would then operate at less than 60 Hz during the fast transient interval. Because the motor's voltage would then proportionally decrease (volts per Hz) with the current peaks, it would allow a voltage drop increase in the motor's conductors. This could allow the motor feeder to be quite long. Chapter 13 illustrates this in a similar application.

A large 460 V induction motor has a torque limit of about 200% of its rated torque. Its horsepower and torque rating are a thermal limit. For the crusher example, not its rms horsepower loading but the torque required during the fast transient loading could size the motor. The motor's frame size would allow its stator's air-gap flux to be sufficient to deliver the crusher's peak torque demand.

## Fast Transient Loading of the Motor's Feeder

The motor feeder, if sized per Table 7.2 and Table B.310.3 in the NEC, 75°C and 447 rms amperes, would be one 500 kcmil/phase. If sized for 400 Hp from NEC Tables 430.15 and 310.16, the required size is two 500 kcmil/phase. This difference shows how the NEC tables are good for only steady-state loads. Comparisons are developed in Chapter 14.

With the motor located a kilometer from the variable-speed drive, the difference between one 500 kcmil and two 500 kcmil feeder sizes is about *a quarter million dollars*. This is illustrated in Chapter 14.

If the motor feeder is 3280 ft long, two elements are important, the required voltage at the motor terminals and the voltage drop in the motor feeder. This is illustrated in Chapter 13.

In the NEC, Table 310.16 cannot be used for 480 V conductor sizing with fast transient loads. In Annex B, Tables B310.1 through 310.10 show that installation methods have a marked effect on a conductor's ampacity. In Table 310.16, many installation methods are all conservatively lumped together. Doing so compromises the temperature rise of a conductor, which is primary to fast transient sizing. Tables B310.1 through 310.10 are valid.

## Fast Transient Loading of Variable-Speed Drives

The drive is not as robust as the motor. It must be rated for the full spike current. Drives have an overload capacity of 150% for only about a half minute, which is not sufficient for the fast transient current spike's duration. At 1000 A for a few minutes and 480 V at 60 Hz, or 830 kVA, the drive must be rated for 940 Hp (i.e., a 1000 Hp rating). Because of the fast transient loading, the motor must be driven by a 1000 Hp drive.

Variable-speed drives that can supply currents in the 2000 A range are comprised of power units connected in parallel. An application of a drive for a large glass furnace used three power units to make up a 4500 A, 950 kVA power supply.

What can also be oversized is the feeder supplying the variable-speed drive. With a 95% drive efficiency, the rated input current for a 1000 Hp drive is about 1050 A. The drive's supply feeder, sizing per 430.22(A) Ex. 2 in the 2002 NEC, in which the drive's nameplate current is multiplied by 1.25, needs an ampacity of 1310 A, even though its design input load current is about 500 rms amperes. Here, per 310.15(C), the required ampacity should be sized with engineering supervision.

## Summary

Machinery having fast transient loads must carefully match the supply, feeder, and motor. Feeders sized by engineering analysis using Table 7.2 can realize significant cost savings. NEC Table 310.16 cannot be used for fast transient cable sizing. Tables B310.1 through B310.10 in Annex B are valid. Other values for 15 kV cables and for European cable sizes can be determined for fast transient loads, like values in Table 7.2, using the methods discussed herein.

# 9

## Introduction to Motor Fast Transient Loading

A fast transient load is a large load that lasts for a few minutes. Spikes of thousands of pound-feet of torque or kiloampere spikes that last a few minutes are examples of fast transient loads. Notice that these loads are neither millisecond transients nor steady-state loads. Chippers that reduce logs to make paper, electric steel furnaces that melt scrap, and starting of large gas turbine generators are some examples of equipment that draw fast transient loads.

In the following chapters, an application is given, the application's motor is selected, its feeder is sized, and its variable-frequency drive is selected. A cost estimate follows. The application is defined as follows.

### A Mining Company Project

A large mining company has made an inquiry. The company's engineering group is designing a machine that is an advancement in processing. The lead engineer wants to know the size of the motor that can operate his machine for 6 min while producing a torque of 1300 lb-ft at 1800 rpm, then followed by operation at 200 Hp and 3580 rpm for 24 min. These two requirements are his worst-case profile. He also adds that 1800 rpm is his best estimate at this time. A variable-frequency drive will allow one motor to meet both loading requirements and can be set at any final rpm value. This loading will be repeated every 30 min for about 10 h/day. Because the motor will drive the shaft of this advanced machine, its shaft's preliminary details can be made available for motor coupling requirements.

The machine will be located at a desert mining site some distance from the processing plant. At the machine's location, a small amount of local power will be required for lighting, small motors, and other miscellaneous loads. There is no limit for site obstructions, and a pole line from the processing plant would be acceptable. Because collisions with the line are possible from trucks, moving cranes, and the like, a line with open conductors on insulators is not desired. Every several months, the machine and equipment will be relocated to another site on the premises, so an equipment skid that mounts the machine, motor, and auxiliaries is planned. The processing plant will be permanently located.

The mining company's design group has asked for a design concept with a preliminary cost estimate. Cost is a factor that can cancel the project; therefore, the estimate must be of a modest cost. Also included with the line's unit cost is any distance limit it may have. A rough estimate is that the cost could exceed half a million dollars. A negotiated bid to prepare this cost estimate was accepted by the mining company.

---

## General Response

Because the processing plant will contain the electric service point for the utility, an electrical room in the plant is assumed that houses distribution equipment. A 480 V three-phase distribution switchgear is assumed in the room, in the form of a lineup of circuit breakers and equipment. The equipment will include a variable-frequency drive for the motor. This drive can power the motor for its two operating modes and be the motor's controller, with disconnecting means in the switchgear. The drive may be included as part of the distribution switchgear if small enough, or be free-standing and located in the room near to the switchgear, which will have the breaker for disconnecting the drive. The source for the motor's outdoor distribution line to the equipment skid will be the drive's terminals.

The line will include the motor feeder and a feeder for the auxiliary load. The feeder for the auxiliary load must be separate from the motor feeder because its loads cannot operate with the motor's lower frequency during the 6 min high-torque intervals. The motor's current will be heavy for the high-torque loading. The auxiliary load is small, so its feeder can be assumed as not limiting for the line's length. A 30 kVA 480-120/208 V transformer will be estimated to feed the auxiliary load.

The motor feeder is required to be insulated, as is the auxiliary feeder. Both can be assumed as jacketed bundled three conductors on a messenger. Both messengers can be suspended from a crossarm. This line arrangement has the same flexible capability for relocating as open conductors would have. Finally, the pole line can include a control-signal cable for the drive's status and operating commands.

The motor will be a totally enclosed fan-cooled NEMA standard type, which is suitable for a mining environment. Its torque requirement isn't special, so a design B speed torque is appropriate. The 200 Hp operation requires a 3600 rpm synchronous speed. The motor will draw a heavy current during its high-torque loading, and this current will be a fast transient load for the feeder's conductors. The feeder voltage drop will be critical for determining the feeder's maximum length.

Because the feeder can be as long as a mile, the drive's output waveform must be of very low harmonic content. The drive must have an inherently low-harmonic inverter of the IGBT type. The drive may also include active input filtering and harmonic control for a harmonic-free connection to the switchgear and service. With line voltage drop a limit, the drive's output voltage capability is also a factor.

In the following chapters, the motor is sized, its line is sized, and the drive is specified. Alternative designs for the motor feeder are also developed. A signaling RTU is included. The cost estimate is then prepared from these specifications and from the auxiliaries' requirements.

# 10

## Fast Transient Motor Analysis

In this section is found the motor elements of a motor operating at 1300 lb-ft of torque at a shaft speed of 1800 rpm. As stated earlier, this motor will not operate at this load continuously. It will operate at this load for 6 min out of every 30 min. For the following 24 min, the motor operates at 200 Hp and 60 Hz. With this operating profile, the 1300 lb-ft can be near the motor's breakdown torque. From a review of a manufacturer's listed motor breakdown torques, a 400 Hp size is selected. This size has a 1470 lb-ft breakdown torque and can produce 1300 lb-ft of torque with a 12% margin.

The motor is a 400 Hp, 460 V, 60 Hz, 3570 rpm TEFC design B motor that conforms to NEMA standards. Its winding is rated for an 80°C rise. This motor will be driven by a variable-frequency drive to provide the operating profile. The first step is to find the motor's electrical characteristics. The next step is to find the thermal characteristics. From the first step, the losses are found, which cause the temperature rises. From the second step, the temperature rises are derived and compared with the motor's temperature limits. NEMA motor standards (NEMA MG-1) define motor temperature limits and restrictions on starting.

### Background

The Roman Empire was based on a slave economy. Today, our economy has the induction motor to do our labor and serve us in so many ways; we take it for granted. If one counted all the induction motors in the world, the total must number in billions. An induction motor consists of a stator with windings in which a rotating magnetic field drags a rotor around with it. The stator is a hollow cylinder with axial slots in its inside surface, in which the windings are installed. The rotor is a squirrel cage of axial bars mounted on its shaft. The stator and rotor are inside a frame having end bells that mount the shaft's bearings.

The induction motor was developed by several men. Its basis is its rotating magnetic field. First theoretically conceived by Tesla, it couldn't be built unless two sources of alternating current were available. At the time, electricity was distributed mostly as DC and single phase. Several voltages and frequencies were in use, including nine frequencies from 25 to 133 Hz for lighting. (Lighting and street cars were the loads. Gaslight was as common as incandescent light.) The two alternating current sources had to be (a) synchronized, (b) equal in voltage, and (c) 90° apart in phase. A voltage at 133 Hz wasn't practical for the induction motor. At a much lower frequency, the required two voltages existed only in the laboratory, generated in two coupled dynamos.

In America, by 1890 two large companies emerged, General Electric and Westinghouse. Three men—Nicola Tesla, trained in Serbia; Charles Scott and Benjamin Lamme, both graduates from Ohio State University—and others worked at Westinghouse to develop the induction motor. Charles Steinmetz also contributed to theoretical models. Steinmetz worked at General Electric. His colleague, G. Kron, was also a contributor. By 1890, electrical knowledge had revolutionized from development by trial and error to development by analytical prediction. But the induction motor needed a standardized voltage system for its production; that is, it required system maturity for its practical universal use. In 1882, only about 70 American universities offered engineering courses. In 1889, more offered engineering, and only 49 offered electrical engineering courses. A standardized system of three-phase 60 Hz was years away.

Voltages were standardized first (circa 1910); frequencies took about 10 more years. Analysis of rotating machines and their characteristics became widely developed and published by about 1920, when standard three-phase 60 Hz power was finally adopted. The giants on whose shoulders we stand today were Dr. C. L. Fortescue, R. D. Evans, and A. C. Monteith of Westinghouse, and others of General Electric. Most of their work was done in the 1910s through the 1930s. The first edition of the Westinghouse-published *Electrical Transmission and Distribution Reference Book*, with 21 authors, was published in 1942. (It is known as the *T and D Book*.) C. F. Wagner was the author of Chapter 6, Machine Characteristics.

The weak link in an electrical system was the motor. With its maturity, the next weak link was the alternator and its prime mover, the steam turbine. Mr. Wagner devotes most of Chapter 6 to the analysis of alternators. He also includes analysis of induction motors. A simplified circuit for an induction motor is shown in Figure 10.1. This figure represents one of the three balanced phases. The load is represented as the rotor resistance, $Rr_s$, multiplied by $(1-S)/S$.

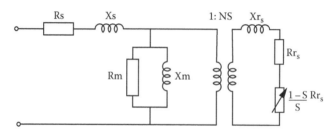

**FIGURE 10.1**
The equivalent circuit of an induction motor.

**TABLE 10.1**

Manufacturer's Data for the 400 Hp Motor

| Load | Shaft (rpm) | Motor (A) | Efficiency (%) | Power Factor | Torque (lb-ft) |
|---|---|---|---|---|---|
| Full load: 400 Hp | 3570 | 421 | 96.2 | 92.6 | 588 |
| ½ load: 200 Hp | 3585 | 220 | 96.3 | 88.3 | 293 |
| Overload | 3466 | 1688 | — | — | 1481 |

*Notes:* Manufacturer's data lists the no-load current as 45 A. The manufacturer also states the stator winding materials are rated for an 80°C rise.

The transformer represents the coupling across the air gap between the stator's rotating magnetic field and the rotor. In this transformer, the stator-side current's frequency is that of the applied voltage, while the rotor-side current's frequency is that of the slip frequency. The inclusion of S with the turns ratio accounts for this difference.

Slip rpm is, if you were riding the shaft, the rate you felt the magnetic field pass you by. Slip value, S, is defined as slip rpm divided by synchronous rpm. In Table 10.1, full-load rpm is 3570. With synchronous rpm of 3600, the slip rpm is 30. S is 30/3600, or 1/120. In the following discussion, the derived load representation is Rr over S. At full load, the load representation is 120 Rr.

Rs and Xs represent the stator resistance and leakage reactance. Rm and Xm represent the excitation's losses and reactance. The applied voltage is the wye value of one phase, that is, phase-to-phase voltage divided by the square root of three. (This circuit is derived in Bose [2002], along with the following circuit.)

---

## The Circuit Used for Analysis

The circuit in Figure 10.1 is a classic circuit, but it is not the simplest form commonly used. The common form, used in the following discussion, is shown in Figure 10.2. Here, rotor value, $Xr_s$, is referred to the stator side of the transformer by the square of the turns ratio and becomes the symbol Xr.

Note that in Figure 10.1, the rotor-side winding resistance, $Rr_s$, divided by $S_s$ can represent both the rotor resistance and the load. In Figure 10.1, one can add the two rotor resistances and find that their sum is thus $Rr/S_s$, representing both the rotor resistance and the load. This value is then transformed to the stator side of the transformer in Figure 10.2 and becomes the symbols Rr and (1 − S)Rr/S.

The displayed element Rm has been excluded. The excitation current is small compared to the stator and rotor currents, about 5%. Most of the no-load current is reactive and flows through Xm. In the discussion that follows,

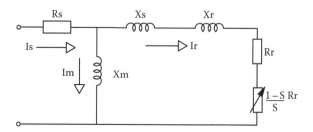

**FIGURE 10.2**
The equivalent circuit used for analysis.

the small amount of real excitation current is neglected. In the following discussion of losses, its value is addressed.

This motor has six phase windings, one pair per phase, with each pair's windings located on opposite sides of the stator. Each pair is arranged in the stator's slots on axes 120° apart. For this arrangement, the convention is that the motor has two poles (two in this circuit showing one of the phases). Most importantly, in the face of each pole, the phase windings make a *sinusoidal* distribution of air-gap flux.

With these six windings and 60 Hz applied frequency, the magnetic field rotates at 3600 rpm. Slower speeds result from adding more windings. Adding six more windings, with 12 windings total (a four-pole motor), a no-load shaft speed of 1800 rpm results. This work keeps it simple, with a six-winding stator as described earlier. Adding more poles to the discussion would be very straightforward.

At rated frequency and voltage, data published by a motor's manufacturer can be used to evaluate some of the circuit constants. The following are listed by a manufacturer for his 400 Hp 3600 rpm motor with 460 V at the motor's terminals. This motor was selected for analysis because the required 1300 lb-ft of torque is about 85% of this motor's breakdown torque.

The shaft rpm column shows that, except for overload, slip rpm is proportional to load for the range from zero to full load. This is true only for small values of slip rpm. As stated earlier, locked rotor current is with a slip speed of 3600 rpm. The coefficient of Rr is then one. But locked rotor current is not used in this analysis.

## Varying Reactance

One other comment about Figure 10.2. The reactance, Xr, is shown as a fixed value. Although many references show the rotor reactance as fixed, when transformed to the stator circuit, this reactance changes with load.

Reactance is the henries multiplied by $2\pi$ and the current's frequency. For motors, the rotor frequency, that is, slip frequency, changes with load. Look at the motor currents and power factors in the manufacturer's data in Table 10.1. At 60 Hz and 400 Hp, the current is 421 A and the power factor is 92.6%. The 265.6 phase volts, 421 A current, and 92.6% power factor yield an input power of 103.5 kW in this one phase.

With this cosine value of 92.6%, *the sine value is 0.378.* The input var value is 42.3 kvar. Adding the two reactances in Figure 10.2, Xtot = Xs + Xr, then I² Xtot = 42.3 Kvar. And Xtot = *0.2382 ohms.* At 200 Hp, the current is 220 A and sine of the power factor's angle is 0.4694, with an input of 27.4 kvar. Here, Xtot is now *0.5667 ohms.* Just by reducing load, the resistance increases, and *so does the reactance.*

Keep in mind that the input current is made up of two currents, the real and the reactive. With the exception of cases at light loads, for changes in load, Xtot changes.

## Torque

Work done in one rotation of a shaft is the driving force, F, times the circumference. In one revolution, F $(2\Pi\ r)$ = ft-lbs

$\tau = Fr$

$2\Pi\ \tau$ = ft-lbs in one revolution, and

$(2\Pi\ \tau)\ (rpm)$ = ft-lbs/min.

With 33,000 ft-lbs/min/horsepower

$2\Pi\ \tau$ rpm/33,000 = Hp, and kilowatts are 0.746 times Hp.

For 588 lb-ft at 3570 rpm, the motor produces an output power value of 400 Hp delivered to the shaft's load. Listed in the manufacturer's table is a value of 588 lb-ft. The torque developed by each balanced phase is 196 lb-ft, and with an output power of 133.3 Hp, or 99.47 kW in each phase.

In Appendix B, the electrical characteristics of this 400 Hp motor are derived from the data in Table 10.1. In Appendix B, a small change to Figure 10.2 is made. The excitation reactance, Xm, is connected to the stator terminals, and Ir flows through the stator resistance, Rs. This simplifies solving for Ir. This change introduces little error, because Ir is almost equal to Is. When calculating stator losses, however, Is flows through Rs. This method is shown in Appendix B.

## Summary

1. Operating the motor for a short time period at 31 Hz and 1061 A produces a torque of 1306 lb-ft. Motor voltage is 248 V.

2. The shaft operates at 1800 rpm and delivers 448 Hp to the load.

3. The power factor is 88.1%, and the efficiency is 82.4%.

4. Total losses for three phases are 71.4 kW.

5. At 60 Hz, 460 V, and 200 Hp, the motor current is 220 A and the power factor is 88.3%.

6. The losses at 200 Hp are 5.73 kW.

Table B.1 lists the results of the calculations in Appendix B.

# 11

## Calculation of the Motor's Temperatures

### Thermal Model

The thermal schematic is shown in Figure 11.1. Using the electrical circuit analogy, thermal resistance is lumped into a fixed value, and thermal capacitance is depicted as an electrical capacitance. Thus, the stator assembly is represented as a thermal capacitance. This capacitance includes the stator steel, the windings, and the rotor because they are all considered isothermal in this model. Between the stator assembly and frame is a lumped resistance. The frame and end bells, considered as isothermal, are represented as a thermal capacitance. Representing the heat transfer from the outside surfaces to the surroundings is a lumped resistance. That this figure resembles that of Figure A.1 is gratifying because the transient heat transfer has already been analyzed in Appendix A.

### Motor Thermal Capacitances

The 400 Hp motor's NEMA frame size is 449T. This motor weighs 3010 lb. This weight may be broken down about as follows in Table 11.1. Also, the specific heat and thermal capacity (weight times specific heat) are listed. The last values have been converted to units of watt-hours per degree Celsius. (Multiply Btu per degree Fahrenheit by 0.5274 to obtain W-h/°C.)

Considering the stator and windings as isothermal (*neglecting the small differences in temperature within the stator assembly), the stator transfers its heat radially to the frame and to the cooling air within the air gap. The stator mounts in the frame with intimate contact so that most of the stator's heat dissipation flows to the frame. In turn, the frame conducts the heat to its cooling fins and end bells. From these surfaces, the heat transfers to the surroundings, through airflow over the fins, and by radiation from the outside surfaces.

---

* NEMA standards state that the temperature rise of the winding hot spot shall be less than 10°C above the winding average temperature.

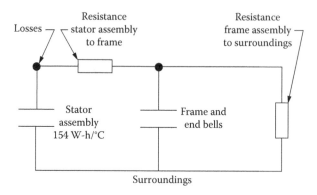

Surroundings

**FIGURE 11.1**
Motor thermal model.

**TABLE 11.1**

Thermal Capacitances

| Element | Weight (lb) | Percent (%) | Specific Heat | Thermal Capacity |
|---|---|---|---|---|
| Frame (cast iron) | 1055 | 35 | 0.20 | 111 |
| End bells (cast iron) | 450 | 15 | 0.20 | 47.5 |
| Stator (silicon steel) | 755 | 25 | 0.22 | 87.5 |
| Windings (copper) | 300 | 10 | 0.09 | 14.2 |
| Rotor (hardened steel) | 450 | 15 | 0.22 | 52.2 |

The rotor transfers its dissipated heat mainly to the stator, through the internal airflow and by radiation. Very few rotor watts of dissipated heat flow through the bearings to the end bells. Because most of the losses are in the stator, and because most of the rotor's dissipation transfers through the internal cooling air to the stator, in this model, all of the losses may be considered in the stator assembly. This is conservative.

This modeled heat flow may therefore consider the stator assembly with rotor as isothermal, and the frame and end bells also isothermal, but colder than the stator. With this model, the rotor is included in the stator and also is isothermal with the stator.

## Temperature Rise Rating

If a motor is operated at 90% of its rated voltage, it will slow down slightly and almost maintain its rated output. The current increases to about 130%. Losses also increase. The power factor decreases. With this 400 Hp motor at full load and rated voltage, its losses are about 12 kW. At 90% voltage, the losses increase 2.5 times. It is this operation that NEMA standards are based on, regarding winding temperature rise. Factory tests under rated full-load conditions show winding temperature rise as less than the winding's 80°C rise capability. Table 11.2 compares rated conditions with operation at 90% of voltage. (Slip and rpm at 90% voltage are found by assuming increases in slip rpm and calculating output. For successive assumed slip, output is seen to rise with more slip, and then fall. The most-output value defines the slip at 90% voltage. This results for a slip of 50 rpm in Table 11.2.)

## Evaluating Thermal Circuit Elements

The thermal resistances are evaluated at steady state, after the motor has heated up. Motor losses in watts (stator and rotor) are assigned to the stator node in the figure.

The sum of the resistors, $R_{total}$, has an 80°C rise with losses of $(P_{in} - P_{out})$. From Table 11.2

$$P_{in} - P_{out} = 29.7 \text{ kW losses}$$

$$R_{total} = 80°C/29,700 \text{ W} = 0.002694°C/W$$

The *apparent* time constant is this resistance (see Chapter 7), 0.002694°C/W, times the thermal capacitance, 154 W-h/°C, of the rotor and stator assembly from Table 11.1:

$$K_{apparent} = (0.002694°C/W)(154 \text{ W-h}/°C) = 0.41 \text{ h, or } 25 \text{ min}$$

With three time constants for a response to steady state, the motor would heat up in about an hour and a quarter. This is typical. And the actual time constant is longer than the apparent time constant, adjusted by Ch, the Coefficient of Henry. (See Chapter 7.)

## Operating Time at 1300 lb-ft of Torque

During the first time constant of an exponential heat-up, the early portion is linear and begins to curve outward with time. This early linear portion lasts for about one-fourth of the first time constant, or 6 min of the time

**TABLE 11.2**

Operation at 460 and 414 V with 400 Hp Output ($X_{tot}$ = 0.2382 Ohms)

| Volts | rpm | Slip (rpm) | Rr/S (Ohms)[a] | R$_{total}$ (Ohms)[a] | cos Φ P.F. | Z (Ohms)[a] | I (A) | kW$_{in}$ | kW$_{out}$ | Losses (kW) |
|-------|------|-----------|----------------|----------------------|------------|-------------|-------|-----------|------------|-------------|
| 460 | 3570 | 30 | 0.55212 | 0.58427 | 0.9260 | 0.63085 | 421.0 | 310.5 | 298.4 | 12.1 |
| 414 | 3550 | 50 | 0.33127 | 0.36802 | 0.8395 | 0.43838 | 545.3 | 326.5 | 296.8 | 29.7 |

[a] Equivalent circuit values.

Note that at 90% voltage (414 V)

- Shaft speed and output drop slightly: 0.5%.
- Power factor drops 90%.
- Motor amperes increase 130%.
- Losses increase 2.5 times.

A further drop in power factor also results from a decrease in Xtot, not in this table, further increasing the current.

mentioned earlier. During these 6 min, the temperature rises at the rate $\Delta t = Q/C$, where Q is the modeled losses and C is as earlier, 154 W-h/°C. For a discussion of this, see Chapter 7.

At 1300 lb-ft of torque, the losses are shown in Table B.1 as 23.8 kW/phase. This is a total loss of 71.4 kW. For the following analysis, this value will be increased from 3.6 kW to 75 kW, or about 6%. This increase adds margin.

Starting with the motor cold, the motor reaches an 80°C rise in

$$\text{Rate of } °C/h = (75,000\,W)/(154\,\text{W-h}/°C) = 487°C/h = 8.1°C/min$$

$$80°C/(8.1°C/min) = 10\,min$$

It is now apparent that this 400 Hp motor, with 248 V at 31 Hz and drawing 1061 A, can be operated for 6 min at a shaft demand of 1300 lb-ft at 1800 rpm. If the winding temperature rise exceeds an 80°C value, it overheats toward the end of the 6 min. Following the 6 min, the motor then cools down. This is not like operation with a continuous 80°C rise. Motor life is affected by winding temperature and the operating time at that temperature.

The temperature rise, increasing from starting temperature rise, can now be found for a time of 6 min. The rate of rise is 8.1°C/min, and the 6 min rise is (8.1°C/min)(6 min) = 49°C.

---

## Cooldown to 200 Hp Operation: Extended Cycle

Cooldown is about the same as heat-up, because the motor only has one thermal time constant. The motor will fully cool down in about an hour and a quarter. Given an operating cycle of 6 min at 1300 lb-ft and 31 Hz, then 75 min (three time constants) at 200 Hp (at 3580 rpm and 60 Hz), the motor will cool down to the temperature rise of 200 Hp operation. Losses at 200 Hp are 5.73 kW. This is from using the values listed in Table 10.1. The stator steady-state temperature rise, with 5.73 kW losses, is

$$\Delta t = Q\,R = (5\,730\,W)(0.002694°C/W) = 15°C$$

The windings will cool down to this value in about three time constants, or 75 min, as given. This is the starting temperature for operating at the 1300 lb-ft load.

## The More Severe Operating Cycle

With a different operating cycle, a hotter winding temperature is reached. Operation demands a cycle of 6 min at 1300 lb-ft of load, followed by 24 min of operation of 200 Hp at 3480 rpm and 460 V. This cycle is repeated for 10 h/day. Let the motor be operated first at 200 Hp for 75 min, achieving a 15°C rise. The temperature rise at the end of 6 min at 1300 lb-ft is, starting at 15°C

$$\Delta t = 15°C + (8.1°C/min)(6 \min) = 15°C + 49°C = 64°C$$

Using the apparent time constant to approximate the cooldown, the formula for exponential decay is

$$\Phi = \theta e^{-t/T}$$

where
  $\Phi$ is the temperature rise at time t
  $\theta$ is the beginning temperature rise
  T is thermal time constant, 25 min
  t is the elapsed time, 24 min

Then with t = 24 min, $e^{-t/T}$ has a value of 0.3829. After 24 min of cooldown, the temperature decays from 64°C, above 15°C, to 0.3829 of the 64°C temperature rise, or 25°C rise, toward 15°C rise. This rise is a 25 + 15, or 40°C rise. This becomes the initial temperature rise for the 6 min heat-up that follows. This 6 min interval has a rise of 49°C above the 40°C beginning, or a peak rise of 89°C. This is followed by a cooldown in the next 24 min to

$$(89 - 15)(0.3829) = 28°C \text{ rise, and } 15°C + 28°C$$

$$\text{or } 43°C \text{ rise for the next beginning rise.}$$

The next cycle begins at 43°C rise and rises to 92°C. Cooldown from 92°C is

$$(92°C - 15)(0.3829) = 29°C, \text{ and } 29 + 15 = 44°C$$

The subsequent cycles approach the cycle of a rise of 49 + 45 = 94°C peak rise, with a cooldown to 30 + 15, or 45°C rise. These cycles have an average value of a 70°C rise. Figure 11.2 illustrates this response.

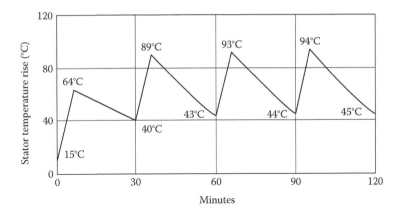

**FIGURE 11.2**
Stator temperature response to the operating profile.

## Afterthought on the Thermal Resistances

In evaluating the thermal circuit elements, the apparent time constant was used to calculate temperatures. This use was because no knowledge of how the total resistance, 0.002694°C/W, was summed from the model's $R_1$ and $R_2$ values. If $R_1$ and $R_2$ were equal, then $R_1$ and $R_2$ would each be 0.001347°C/W, and Equation A.13 could be used to calculate the motor's time constant. If so, the result would be a time constant of 33 min, longer than the 25 min of the apparent time constant, and the value of Ch would be 33/25, or 1.32 and less than 2. With a longer time constant, the peak temperature rise would be less than 94°C.

Comparing $R_1$ and $R_2$, the frame's outside surface area is greater than the stator's surface area within the frame, implying that $R_2$ might be less than $R_1$. Choosing an arbitrary ratio for $R_1/R_2$ of 1.333, and from using these values of $R_1$ and $R_2$ in (A.13) and (7.4), the following results:

$R_1 = 0.001539°C/W$

$R_2 = 0.001155°C/W$

$Ch = 1.2375$

$K' = 31$ min

$\Theta_6 = 41°C$ rise after 6 min, instead of a 49°C rise using the apparent time constant

Using this arbitrary set of thermal resistances in this crude thermal model implies that the plotted temperature rises in Figure 11.2 are hotter than actual values might prove. If so, these temperature rises are conservative.

## Summary

The 400 Hp motor has been modeled as an isothermal stator–rotor assembly transferring its dissipated heat to the frame and end bells. The frame and end bells are also modeled as isothermal and are cooler than the stator assembly. The frame and end bells transfer the dissipation to the motor's surroundings. The thermal capacitance of the stator assembly is its composite weights and specific heats.

The thermal resistance has been calculated between the windings and the motor's surroundings. The resistance is based on the motor's rated 80°C rise and losses when, per NEMA standards, operating at 90% voltage.

For an operating cycle with the 400 Hp motor, of 6 min at 1300 lb-ft of torque and 31 Hz, followed by 24 min at 200 Hp and 60 Hz, the modeled stator's temperature response has been calculated. The results are approximate. The response is shown in Figure 11.2. The windings operate at a 94°C rise for a peak each cycle. The windings cool to 45°C to begin the next cycle. The average rise is 70°C rise for each cycle.

The apparent time constant of the motor's exponential thermal response has been approximated as the product of the stator assembly's capacitance and the thermal resistances between the windings and the surroundings. A refinement of the time constant, discussed in Chapter 7, employing Ch, could not be used. Some method to evaluate the two separate thermal resistances in the model, Figure 11.1, would be necessary to apply Ch. The 6 min heat-up of the stator assembly is linear, but the apparent time constant dictates the following 24 min of cooldown.

When a motor is operated at 90% voltage and rated load, its losses greatly increase from its losses at full-voltage operation; and its stator temperature rises to its maximum operating rise. NEMA standards base the stator's rated temperature rise on this operation.

# 12

## Motor Conclusions

### Frame Size

In the introduction chapter, Chapter 9, the requirements for a motor's shaft load was defined as a demand needed by the machine the motor was to drive. The demanded cycle had a 30 min period, with a required torque of 1300 lb-ft at 1800 rpm for 6 min, followed by 24 min at 200 Hp and 3585 rpm. The analysis shows that one manufacturer's TEFC 449T frame size can house a stator and rotor that can deliver the torque.

This 449 T frame size can house stators and rotors over the range of horsepower ratings from 250 to 450 Hp. With this 400 Hp combination, the average winding temperature is a 70°C rise, which is below the motor's rating of an 80°C rise. However, the peak temperature is about a 94°C rise. The windings could see a repeated thermal expansion to a size bigger than their rated size, as well as the repeated expand–contract stress. These *results are preliminary, and may be on the low side. (The thermal model is crude.)*

An alternative, if possible, is to lengthen the 24 min cooldown period. For example, if the cooldown interval is extended to 30 min, instead of the 24 min interval, the peak rise becomes 87°C, and the profile has an average rise of 63°C. Another alternative would be to shorten the 6 min operation. This too would moderate the temperature swing.

Not only is this high-stress winding operation a concern: is the rotor's shaft diameter adequate for the high torque, and are its bearings adequate for the cycling stress? The rotor's shaft diameter should match the machine's driven-shaft diameter, and the coupling should be adequate for the cyclic load. The rotor's bearings should also compare with the driven shaft's bearing types and sizes.

These added considerations may result in the next larger frame size being used for the motor, a 509 or 5011 frame with a matching rotor shaft diameter. A physically larger motor, with less losses and more thermal capacitance, would provide better design and wear performance.

The next sizes of frame, a 509 or 5011, will increase the motor weight from about 3100 to 4000 lb. The cost increase is about from $51,000 to $60,000 for the motor.

## Purchasing Requirements

The motor requisition must include that the supplier furnish the following data points:

- At 460 V, full load and 200 Hp operating data.
- No-load current at winding rated temperature.
- Average hot stator resistance of the stator phase windings.
- Operating data at an overload near breakdown, and at breakdown, at rated voltage, and 60 Hz.
- Equivalent-circuit element values for these conditions, with average frame temperature.
- The data must justify calculations that comply with the electrical analysis in Appendix B.
- Motor component weights, like the data in Table 11.1.

The supplier shall perform a constant-load procedure at motor rated load lasting to steady state (2 h). This procedure is justified to verify the predicted thermal response time and peak temperature profile. Temperature detectors shall be embedded with the windings. Thermocouples shall be applied to the frame at several locations, and to the end bells, for recording temperatures during this test. These thermocouples and embedded sensors will be used to record the temperature response versus test time of a cold motor with a constant load equal to the motor's rating. This test would be done with 90% voltage. All electrical values and response temperatures shall be recorded and reported.

Measurements will be used to evaluate elements in the motor and the thermal models. With different average temperatures for the stator, frame, and end bells

1. Measured response time of the stator can be evaluated.
2. Values for the thermal resistances in Figure 11.1 can be evaluated.
3. The time constant can be evaluated.

(Evaluating the time constant assumes that fairly accurate thermal capacitances from component weights are obtained.)

Because this application is for a new process, keep in mind that available motor data was probably collected decades ago and has been republished more than once. Errors could have crept into the available data.

The motor is to be used in a new process. If this process requires a narrow tolerance on operating rpm, a shaft rpm sensor to feed back to the variable-speed drive is justified. Also, requirements dictated by matching the motor's shaft to its driven shaft (diameter, coupling, bearings, etc.) must be included in the purchasing requirements.

Another requirement to be resolved in the purchase contract will be the supplier's guarantee for this new application of his motor. His natural caution will lead to limit his guarantee of the product. A sheared shaft or premature bearing failure, a winding failure, etc., will not be expected to be part of his guarantee. A proper design of the driven shaft, its coupling, and its bearings cannot be assumed. Design analysis of these elements must be a part of the review by all parties.

# 13

## Line Design

After voltage drop, there are four elements to designing the line, the load, sizing the cable, ampacity, and cable temperature. The load is based on the motor. Ampacity is that which is adequate for the rms value of the motor current profile plus the fast transient spike. Choosing cable temperature includes the choice of ambient temperature for the line design.

### The Load

The load is the current profile summarized in the motor electrical analysis. It is repeated here. The peak current is 1062 A, which lasts for 6 min. This is followed by 24 min with a current of 220 A. This cycle is repeated for 10 h/day.

The rms value of this current profile is

$$(1,062)^2 (6\min) = 6,767,064$$

$$(220)^2 (24\min) = \frac{1,161,600}{7,928,664}$$

Dividing this sum by 30 min yields the mean square, or 264,289 A². The rms value is the square root, or 514 A.

This is an rms load of

$$(480)(514)(1.732)\,10^{-3} = 427\,kVA$$

Note that the 514 A value slightly exceeds the motor's current, 421 A, when operating at 400 Hp and 460 V. But it doesn't exceed the motor current, 547 A, when operating at 90% voltage and 80°C rise. Also, the resulting motor temperature profile and 70°C average temperature are noted and discussed in the "Motor Conclusions," Chapter 12.

## Cable Ampacity and Size

There are two methods in the *National Electrical Code* for sizing a 460 V motor feeder. The first method is based on the motor horsepower. One looks up the motor's full-load current in Table 430.250, multiplies this current by 1.25, and with corrections, finds the conductor size in Table 310.15(B)(16).

The other method is given in Paragraph 310.15(C) that permits calculating the square of ampacity as a function of conductor temperature rise, AC resistance, and a fixed value of the conductor's effective thermal resistance. These two methods are compared in Table 14.3, for limits of line length, voltage drop, and cost. The method used herein is the calculation method of 310.15(C). This method is developed in *Chapter 8*.

The cable is to consist of three copper conductors suspended with a messenger on poles. *Table 310.15(B)(20)* in the *National Electrical Code* lists current ratings for conductors suspended with a messenger. Because the cables are carrying fast transient loads, the conductor insulation is chosen as a rubber type (RHW), with a jacket that is sunlight resistant. These materials, having high thermal capacitance, provide a minimum temperature rise in responding to the peak current. The insulation and jacket are two thirds of the conductor's thermal capacitance. This is discussed in *Chapter 7*. Also, the line can be subject to collisions with cranes and the like. The jacket helps protect the conductors from these hazards.

Table 310.15(B)(20) lists current values for messenger-suspended cables. Conductor sizes and temperature ratings of 75°C and 90°C are listed. The design ambient temperature for the line is 40°C, reflecting the desert location for the line. This includes adjustments. The conductor rise is thus 75 less 40, or 35°C.

Ampacities for a conservative 75°C conductor rating will be used. For a current of 514 A, a 600 kcmil size is rated at 553 A. The temperature rise with 514 A rms is

$$(514/553)^2 \, 35°\text{C} = 30°\text{C rise}$$

This leaves only a 5°C rise for the current spike, which isn't enough. Then a size in the 700 kcmil range for a single-conductor line is required. A more costly line will be with two conductors per phase, that is, two three-conductor cables, each carrying 514/2 A. This motor feeder will then consist of two three-conductor bundles with messengers. Each bundle is to carry 257 A rms current, with spikes of 1062/2 A, or 531 A.

Table 310.15(B)(20) lists, for a 250 kcmil size, a value of 320 A, and for a 350 kcmil size, 359 A.

With a 250 kcmil size, $(257/320)^2 \, 35°\text{C} = 23°\text{C}$, leaving a 12°C rise for the spike.

With a 350 kcmil size, $(257/359)^2 \, 35°\text{C} = 18°\text{C}$, leaving a 17°C rise for the spike.

The temperature rise due to the current spike is found in Table 7.5, *480 Volt Cable Heating Tabulation.* Listed in this table are spike values for cable types and sizes. With two bundles, the current spike is 1062/2 = 531 A. The pertinent table data are the 600 V rated spike values for the earlier mentioned cable sizes. These are the entries for 1000 A for 5 min. These values can then be interpolated as discussed in Chapter 8.

Interpolating

$(531 \text{ A}/1000 \text{ A})^2 = 0.282$

$6 \text{ min}/5 \text{ min} = 1.2$

$(1.2)(0.282) = 0.338$

Referring to Table 7.5 for the spike, 1000 A for 5 min yields a rise of

With a 250 kcmil size, 45.6°C, and $(0.338)(45.6) = 15°C$ spike

With a 350 kcmil size, 27.0°C, and $(0.338)(27.0) = 9°C$ spike

---

## Cable Temperature

Then with 250 kcmil, the rms rise plus the *peak rise* is 23 + 15 = 38°C rise, and peak cable temperature is 40 + 38 = 78°C.

Then with 350 kcmil, the rms rise plus the peak rise is 18 + 9 = 27°C rise, and cable temperature is 40 + 27 = 67°C.

With the cable rated for 95°C, either size will be suitable. The remaining requirements are the voltage drop and the line length. A 350 kcmil cable has slightly less impedance than the 250 kcmil cable. For voltage drop, the impedances listed in Chapter 9 Table 9 will be used.

---

## Voltage Drop

Complex algebra is employed to calculate the voltage drop. Two conditions are considered: the motor current during the reduced frequency operation during the spike and the motor current during the remaining 24 min of the cycle operating at 200 Hp. Table B.1 lists the data for the spike, and Table 10.1 lists the 200 Hp data.

All voltages in voltage drop calculations are phase to neutral. The impedances listed in Table 9 are phase to neutral for 60 Hz. During the 60 Hz–200 Hp operation, the source voltage at the terminals of the variable-frequency drive is higher than nominal, 290 V, which is 5% higher than 277 V nominal value. (The drive's standard output voltage range is at least ±5%.) The motor voltages are the values derived in the motor analysis, Chapter 10. During 60 Hz operation, the motor voltage is 265.6 V. During 31 Hz operation, the motor voltage is 141 V, and the drive terminal voltage is 277 V.

Values of voltage drop are calculated per 1000 ft of cable length. Impedance values in Table 9 of the NEC are listed this way. Impedance values in Table 9 are for PVC conduit entries. Proximity effects in a nonmetallic conduit are the same as in air. The cable impedances are listed in their complex form. The motor currents discussed earlier are in their polar form. Using their values of power factor, they also are given in their complex form, as derived in the Appendix B motor analysis.

Voltage drop calculated for the 250 kcmil cable and the 60 Hz operation is as follows.

With two conductors per phase, voltage drop $\equiv \Delta V$

$$I_{60\,Hz} = 110\ @ -27.8°\ A = (97.7 - j51.6)\,A$$

$$Z_{60Hz} = (0.052 + j0.041)\ \text{from Table 9}$$

$$\Delta V = (97.7 - j51.6)(0.052 + j0.041) = (5.08 + 2.12) + j(4.01 - 2.68) = 7.20 + j1.33$$

$$\Delta V = 7.32\ @ 10.5°\ V\,/\,1000\,ft$$

## Line Length

The line length is then 24.4 V allowable voltage drop, and a length of 24.4/7.32 = 3330 ft. This is broken down as 5% of the length in each end's local routing and 3000 ft on the pole line. Table 13.1 lists the 3000 ft on the pole line. In the pricing section, *Chapter 14*, the conductor purchase length is the 3330 ft increased by 10%, or 3670 ft/conductor. The line lengths in the table show that the 200 Hp operation limits the line length to 3000 ft with 250 kcmil size, and 3600 ft with 350 kcmil size.

## Auxiliary Power Feeder Size and Signal Cable

The load is estimated as a design value of 30 kVA. A 480 to 208/120 V transformer with secondary feeding a panelboard is assumed, both located on the skid near the motor. The design load current on the 480 V transformer's

**TABLE 13.1**

Voltage Drop and Line Length

|  | Motor Voltage | Source Voltage[a] | Difference (ΔV) | Conductor Current | Conductor (Ohms/1000 ft) | Voltage Drop (V/1000 ft) | Line (ft) |
|---|---|---|---|---|---|---|---|
|  |  |  |  | 250 kcmil conductor |  |  |  |
| Spike | 141.1 | 277 | 135.9 | (461 − j 263.1) | 0.052 + j 0.021 | 29.8 @ 7.7° | 4000 |
| 200 Hp | 265.6 | 290 | 24.4 | (97.7 − j51.6) | 0.052 + j 0.041 | 7.32 @ 10.7° | 3000 |
|  |  |  |  | 350 kcmil conductor |  |  |  |
| Spike | 141.1 | 277 | 135.9 | (461 − j 263.1) | 0.038 + j 0.021 | 23.0 @ −1° | 5300 |
| 200 Hp | 265.6 | 290 | 24.4 | (97.7 − j51.6) | 0.038 + j 0.040 | 6.09 @ 18.7° | 3600 |

[a] Source is the terminals of the variable-frequency drive.

primary is 36 A. The primary will use a 10% tap below rated tap. With this tap, an allowed line drop of 27.7 V is created. The 10% is a −5% tap on the high-voltage side and a +5% tap on the low-voltage side.

The needed line impedance is then 27.7 V/36 A, equals 0.769 ohms, or for 4000 ft of conductor, 0.192 ohms/1000 ft. An AWG 2 conductor size has an impedance value of 0.19 ohms/1000 ft. Hence, AWG 2 will supply this load.

The skid's signals and alarms are estimated to number 15. Then a cable containing 20 shielded pairs of AWG 20 wires, with overall shield, will be assumed for the signal cable. These shields will provide signal isolation, but further isolation will also require spacing away from the 480 V power cables on the line.

## Alternative Conductor Sizes

The line cost with one conductor per phase will require a 750 kcmil size. This cable is rated 638 A, and rms temperature plus spike puts the conductor at 73.5°C, which is less that the 75°C design. The voltage drop at 200 Hp load, 9.4 V/1000 ft, allows a length of 2340 ft for the line. This is the shortest line length yet considered, but the required conductor length to be purchased is only 7720 ft. The conductor length to be purchased for the 250 kcmil size is 22,000 ft, for the 3000 ft line length.

If the common method is used to size the motor's conductor, Table 430.250 lists the full-load current for a 400 Hp motor as 477 A. This is multiplied by 1.25 for ampacity requirements, or 596 A. Entering ampacity Table 310.15(B)(16) in the 75°C column, two 500 kcmil are sized. On a messenger, this size is rated 638 A. With this size, the added spike temperature rise total is less than the 75°C rating. The allowable line length, however, is increased to 4200 ft. A comparison is made with the 350 kcmil size. The line length of 3600 ft is used for a price comparison.

## Transmission at 2400 V

It should be noted that the line's transformer source voltage is 2400 V, while load voltage is 4% less, 2300 V. Costs totaled for the earlier mentioned 480 V conductors, the line construction, and the motor show that the conductors' costs are very high. An alternative of a three-phase line, stepped up at the source to 2400 V and feeding the motor at 2300 V, dramatically reduces conductor cost. Although this makes the motor more costly and requires adding the transformer, the conductor savings more than pay for the increase of transformer and added motor cost. This is because

1. The transformer costs much less than the earlier mentioned conductors.
2. For the motor's feeder, the three 2300 V conductors are small.
3. The conductor voltage drop at 2300 V is small for the same line lengths mentioned earlier.

This offsets the added transformer impedance.

Also, the cost of the 2400 V motor, compared with a 460 V motor, is much greater than a step-down transformer's cost. This second transformer's impedance is also small.

In Table 13.1, line distances are worked out based on the conductor impedances, with voltage drop allowances. With these same allowances, the conductors' and transformers' impedances are equivalently developed. For the motor feeder, transformer impedance is included with cable impedance. A 500 kVA size transformer is sufficient for the motor's rms load of 427 kVA. This oil-filled transformer will not be stressed by the spike. This 500 kVA size has an impedance of 5.0%.

The need here is for the transformer's complex impedance, so that the transformer and cable can be combined. It can be shown that the resistive part of the transformer's complex impedance is the full-load loss as a percent of the size. This should be only the loss in the windings, but because the no-load magnetizing loss is small, the total loss at full load is accurate for use here. The full-load loss as a percent of kVA is the resistive loss in percent. This loss is (100% less efficiency in percent).

Standard source voltage is 2400 V; standard motor voltage is 2300 V. Per-phase values are 1386 V and 1328 V.

The 500 kVA transformer with 5.0% impedance has an efficiency of 98.5%. Loss as a percent of kVA is (100% − 98.5%) = 1.5%, and

$$Z\% = 5.0\% = 1.50\% + j4.77\%$$

With losses as a percent of kVA, the transformer's impedance on the 2.4 kV (1.386 kV phase) side at full load is $I_{FL}$ = 500 kVA/(2.400 kV)(1.732) = 120.28 A.

$$Z_{FL} = 1386\,V/120.28\,A = 11.52\,\text{full-load phase ohms}$$

$$(11.52)(0.050) = 0.576\,\text{transformer phase ohms}$$

Transformer impedance is

$$(11.52)(0.0150) = 0.1728 \, \text{ohms}$$

$$(11.52)(0.0477) = j0.5495 \, \text{ohms}$$

$$Z_{xfmr} = 0.1728 + j0.5495 = 0.576 \, @ \, 72.54° \, \text{phase ohms}$$

For two transformers, $Z_{xfmrs} = 0.3456 + j1.100 = 1.15 \, @ \, 72.54°$ phase ohms on the 2400 V side.

The line current is for a load of 200 Hp at 2300 V. At 200 Hp = 149.2 kW (149.2/3 per phase, or 49.73 kW), the kVA is, with a PF of 0.926 and efficiency of 0.964 (see Chapter 10)

$$kVA = 49.73 \, \text{kW} / (0.926)(0.964) = 55.71 \, \text{kVA load/phase}$$

The motor current is

$$55.71 \, \text{kVA} / 1.328 \, \text{kV} = 41.95 \, \text{A at 200 Hp}$$

The motor power factor is 92.6%, $\cos \varphi = 0.926$, $\sin \varphi = 0.378$.
Thus, 41.95 A = 38.85 − j15.84 A.
The two transformers' voltage drop is I Z:

$$(38.85 − j15.84)(0.3456 + j1.100) = (13.43 + 17.42) + j(42.74 − 5.47)$$

$$\text{Voltage drop} = 30.85 + j37.27 \, \text{V/phase}$$

The allowable voltage drop at 480 V is source voltage minus motor voltage. But with the transformer 5% taps, more voltage drop is allowed. The transformer voltage drop can be offset by 10%. And

$$\text{Voltage drop} = 0.9(30.85 + j37.27) = 27.77 + j33.54 \, \text{V/phase}$$

Because the taps allowance has already been included, the turns ratio of 480/2400 can be used. Working on the 480 V motor side, these values become

$$(27.77 + j33.54)(480/2400) = 5.55 + j6.71 \, \text{phase volts}$$

$$5.55 + j6.71 = 8.71 \, \text{V}$$

The allowable voltage drop is 290 less 265.6, or 24.4 V. Less the transformers

$$24.4 − 8.7 = 16 \, \text{V left for the line conductor drop}$$

The 5 kV conductor size is selected as AWG 2. This size has more than adequate ampacity for the 41.9 A motor current, as well as the rms current and spike. This conductor is manufactured by most cable suppliers and can be found at usual prices. The impedance is

$$Z\text{cable} = 0.19 + j0.37 \text{ ohms}/1000\,\text{ft}$$

The voltage drop of 3600 ft of cable is

$$(38.85 - j15.84\,\text{A})(0.684 + j1.332\,\text{ohms}) = (26.57 + 21.10) + j(51.75 - 10.83)$$

2.4 kV voltage drop = 47.67 + j40.92 V, and for the 480 V side, the drop is 9.53 + j8.18. At 480 V, the voltage drops are

Transformer = 5.55 + j6.71
Cable = 9.53 + j8.18
15.08 + j14.89 = 21.2 V, which is <24.4 V

This leaves 3.2 V to be dropped with 480 V in the process building and 460 V at the skid. With short lengths of 230 kcmil cables installed in the process building and on the skid, this is an acceptable allowance for these short lengths.

The said line combinations are entered in the pricing tables of Chapter 14.

# 14

## Comparisons of Line Alternative Costs

### Cable Prices for 600 V Lines

There are three cables to be priced: a motor feeder, an auxiliary power feeder (both 600 V class), and a signals cable. All are on a pole line suspended by messengers. Each messenger for the power cables suspends three-phase single conductors, insulated and jacketed. Of the motor feeder, all but one alternative are two messengers per feeder. The 750 kcmil feeder is one messenger per feeder.

Each 600 V power conductor is class B stranded copper with RHW insulation rated 90°C for wet or dry use. The jacket is of sunlight-resistant polyethylene or equal material. Insulation thickness, 95 mils; jacket thickness, 80 mils.

### Motor Feeder

This feeder's cables have four alternatives at 480 V:

1. A 250 kcmil 21,000 ft length of conductor (two conductors per phase)
2. A 350 kcmil 25,200 ft length of conductor (two conductors per phase)
3. A 500 kcmil 25,200 ft length of conductor (two conductors per phase)
4. A 750 kcmil 14,000 ft length of conductor (one conductor per phase)

With a 250 kcmil conductor, the feeder is limited to 3000 ft. With a 350 kcmil conductor, the feeder is limited to 3600 ft. Impedance and allowed voltage drop limit the length. The maximum lengths are defined in this study.

Another alternative is a 2400 V line with one conductor per phase. This cost is also developed.

## Aux Power Feeder

This feeder's cables are 600 V AWG 2, 12,600 ft length (one conductor per phase).

## Signals Cable

This cable is 300 V rated, having multiple 20 AWG individual shielded pairs, with overall shield and sunlight-resistant jacket. The cable is also tray-cable rated for routing inside the process building. The number of pairs is 20. The signals cable is on its own messenger, separated from the power conductors by 1 ft.

## Galvanized Steel Messenger Cable and Spiral Wrapping Tape

Alternatives of either three or four messengers are required. The steel messenger cable is seven-strand. Each strand is 0.170 diameter, 11,500 lb strand breaking strength. Overall diameter is 3/16 in. Cost is 0.63/ft.

The tape comes in coils of galvanized sheet steel 3/4 in. wide, 3 ft length per foot of messenger. The cost is estimated at twice the steel cable: $1.25/ft of messenger. Material cost of messenger and tape is 1.88/ft (Tables 14.1 through 14.3).

| Pole and Messenger Costs | |
|---|---|
| Materials: Messenger cable, coil tape, and hardware 1.88/ft | |
| Weatherheads, 2/line | 85.00 |
| Poles, fully equipped | 450.00/pole |
| Pole grounds | 100.00/pole |
| Total 550.00/pole | |
| Labor: Per material cost, twice material cost | |
| Install messengers: Messengers and tapes per foot 3.76 | |

**TABLE 14.1**

Alternative 480 V Costs for Messengers and Poles

| Alternatives | Material | Labor | Total |
|---|---|---|---|
| 250 kcmil 3000 ft | | | |
| Messengers (4) | 22,560 | 45,120 | 67,680 |
| Poles 16 | 8,800 | 17,600 | 26,400 |
| | | | 94,080 |
| 350 kcmil 3600 ft | | | |
| Messengers (4) | 27,060 | 45,120 | 72,180 |
| Poles 19 | 10,450 | 17,600 | 28,050 |
| | | | 100,230 |
| 500 kcmil 3600 ft | Same | | 100,230 |
| 750 kcmil 2340 ft | | | |
| Messengers (3) | 11,730 | 23,460 | 35,190 |
| Poles 13 | 7,150 | 14,300 | 21,450 |
| | | | 56,640 |

**TABLE 14.2**

Alternative Cable Costs

| | | | |
|---|---|---|---|
| 250 kcmil 3000 ft | 56,260 | 30,000 | 86,260 |
| Aux and signals | | | 10,230 |
| | | | 96,490 |
| 350 kcmil 3600 ft | 94,200 | 43,200 | 137,400 |
| Aux and signals | | | 12,280 |
| | | | 149,680 |
| 500 kcmil 3600 ft | 131,800 | 64,800 | 196,600 |
| Aux and signals | | | 12,280 |
| | [This is the NEC motor feeder method.] | | 208,880 |
| 750 kcmil 2580 ft | 93,170 | 54,180 | 147,350 |
| Aux and signals | | | 8,800 |
| | | | 156,150 |

## Alternative 2400 V Line

Instead of a 600 V line, a step-up transformer at the process building and a step-down transformer at the motor skid will reduce the motor feeder's cost. The length limit with these transformers is 3600 ft (Table 14.4).

**TABLE 14.3**

Total Alternative 480 V Line Costs

| 250 kcmil | | Poles and hardware | 94,080 |
|---|---|---|---|
| | | Cable | 96,490 |
| | 3000 ft | 63.52/ft | 190,570 |
| 350 kcmil | | Poles and hardware | 100,230 |
| | | Cable | 149,680 |
| | 3600 ft | 69.42/ft | 249,910 |
| 500 kcmil | | Poles and hardware | 100,230 |
| | | Cable | 208,880 |
| | 3600 ft | 85.86/ft | 309,110 (NEC method) |
| 750 kcmil | | Poles and hardware | 56,640 |
| | | Cable | 156,150 |
| | 2580 ft | 82.48/ft | 212,790 |

**TABLE 14.4**

Total Alternative Line Cost for a 2400 V Line

| Poles and hardware | 79,030 |
|---|---|
| Two 500 kVA transformers | 25,000 |
| 5 kV cable | 6,290 |
| Aux and signals cables | 12,280 |
| 3600 ft: 27.11/ft | 122,600 |

## Summary

1. The tables show that with 250 kcmil motor conductors, the length is limited to 3000 ft. The line cost is $190,700. Using a 350 kcmil conductor adds $59,000 and extends the length 600 ft.

2. Sizing the motor feeder by conservative engineering calculation yields a 350 kcmil conductor and costs $250,000.

3. Sizing the conductor as a motor feeder based on 400 Hp yields a 500 kcmil conductor per the code-tables method and adds about $60,000 to the 350 kcmil cost.

4. Basing the design on one 750 kcmil size lessens cost but shortens the maximum length.

5. Using the 5 kV design cuts the line cost in half. The line length is voltage drop limited to 3600 ft, and the cost is $122,600. The cost per foot is that of the poles, messenger, and conductors, excluding transformers. With 200 ft span per pole, the cost is $27.11/ft.

6. At each end of the line in the process plant and on the skid are conductor and installation costs. These costs are estimated as 5% of the 250 kcmil line cost, $10,000. This adds to the 5 kV line cost, for $132,600.

7. Voltage drop limits the line length to 3600 ft. If this maximum is insufficient, the length can be extended by using a larger cable size. This will lower the value of impedance, and thereby will extend the limit of line length.

8. The motor weighs 4400 lb. A 500 kVA oil-filled transformer with air terminal chambers weighs 4600 lb. The transformer's outdoor mounting on the machinery skid may be a problem.

# 15

## Preliminary Cost Estimate

The estimate's elements are process plant costs, line costs, and skid costs. Process plant costs are the circuit breaker supplying the variable-frequency drive and the drive. (A PLC on the skid is priced in the following; it could be in the process building.) Cable costs and installation are in the line estimate. Skid costs are the motor, excluding its mounting and coupling; a 30 kVA 120/208 V transformer; and panelboard.

### The Process Plant Cost

The variable-frequency drive and circuit breaker constitute this cost. The drive is rated for the peak current. This current lasts for 6 min. For a drive, 6 min is a continuous-load interval. Thus, the drive is rated for 1200 A at 480 V. Its installed cost is estimated at $37,100:

| | |
|---|---|
| Base price | 25,000 |
| Added features | <u>7,255</u> |
| | 32,255: Install at 0.15 times    Drive cost $37,100 |

The circuit breaker is assumed in a switchgear lineup. The spike current is 1060 A on the drive's output at a voltage of 248 V. This a dc power of 262.9 kW that appears on the dc bus. The input to the rectifier is this power divided by drive efficiency, which is 95%, and at 480 V. This is an input power of 277 kW at 480 V. Hence, peak input current is 333 A. The next frame size is 800 A. Its short-time and long-time trip settings must pass 340 A for 6 min.

| | | | |
|---|---|---|---|
| Breaker base price | 17,000 | | |
| Added features | <u>400</u> | | |
| | 17,400 Install at 600 | Breaker cost | $18,000 |
| | | Total plant cost | $55,100 |

### The Line Cost

The line cost estimate is given in the summary of Chapter 14, $132,600.

## Skid Costs

A comparison of the 2300 V motor cost and the cost of the 450 Hp 460 V motor with step-down transformer was discussed in the line analysis, Chapter 13. These list prices are given in Table 15.1.

The 400 Hp motor's torque has no margin for the required 1300 lb-ft torque. The 450 Hp motor has an adequate margin for the torque, lower losses, and a heavier frame, yielding cooler peak stator temperatures. (A manufacturer's motors are priced above that and are different from another manufacturer's 400 Hp motor used in the motor analysis of Chapter 10. See Chapter 12 "Motor Conclusions" for a discussion.)

The 450 Hp 460 V motor fulfills the analytical requirements. It is used in this estimate:

| | |
|---|---|
| 450 Hp motor, installed (its transformer cost is included in the line's cost) | $65,000 |
| Also with the motor is the 30 kVA transformer and panelboard. Their cost is | $3,900 |
| A small PLC installed cost is | $8,600 |
| Total skid cost | $77,500 |

**TABLE 15.1**

Costs of a 450 Hp Motor with a 2300 V or 460 V Winding

| | |
|---|---|
| 450 Hp 2,300 V motor | $93,500 |
| 450 Hp 460 V motor | $60,580 |
| 2400 – 480 V 500 kVA transformer | 8,750 |
| | $69,300 |
| The difference is a savings of $24,200. | |
| The motor cost is from | |
| 400 Hp 460 V TEFC breakdown 1,350 lb-ft: list price | $51,490 |
| Install @ 7% | 3,510 |
| | $55,000 |
| 450 Hp 460 V TEFC breakdown: 1518 lb-ft: list price | $60,580 |
| Install @ 7% | 4,420 |
| | $65,000 |

## Summary (Table 15.2)

**TABLE 15.2**

Estimated Total Cost and Breakdown

| | | |
|---|---|---|
| Process plant equipment | | $55,100 |
| Line 3600 ft (1.1 km) | | $132,600 |
| Skid | | 77,500 |
| Material and labor | | $265,200 |
| Freight and taxes 10% | | 26,500 |
| | | 291,700 |
| Design | 25% | |
| Management | 15% | |
| | 40% | $116,700 |
| | | $408,400 |
| Contingency 10% | | $41,600 |
| Preliminary estimate total | | $450,000 |

*Notes:* 1. The line cost mentioned earlier accounts for 50% of the material and labor cost. The cable costs with 250 kcmil cable would add $100 thousand to the estimate. The cable costs with 350 kcmil cable would add $200 thousand to the estimate.

2. If the NEC motor full-load amperes method is used, requiring 500 kcmil cable, the added cable costs to the estimate would be $300 thousand.

3. The design cost is based on the design hours estimated for the project.

## Costs per Foot (Table 15.3)

**TABLE 15.3**

Alternative Costs per Foot

| Size | Max Length | Dollars per Foot |
|---|---|---|
| | 480 V | |
| 250 kcmil | 3000 ft | 63.52/ft |
| 350 kcmil | 3600 ft | 69.42/ft |
| 500 kcmil | 3600 ft | 85.86/ft |
| 750 kcmil | 2580 ft | 82.48/ft |
| | 2400 V | |
| 2 AWG | 3600 ft | 27.11/ft (in estimate) |

# Appendix A: Derivation of Equations for Transient Heat Flow from Cables in Conduit

The model for the derivation is shown in Figure A.1. In this model, the three conductors are considered as spaced together and located in the axial center of the conduit. The conduit is assumed to be suspended in free air. The conduit and cables are regarded as long enough to have no axial heat flow, and a unit length of the conduit and cables is represented. Thereby, the heat flow from the conductors is more or less only radial to the inside conduit wall and to the surrounding air. This simplifies the heat flow as one-dimensional. Also, any temperature differences among the three conductors and their coverings are ignored and assumed as negligible, and the small radial temperature difference through the conduit wall is neglected. The conductors are represented in Figure A.1 as the left node.

The thermal resistances in this model are assumed constant and do not change with temperature. This is not strictly true, but the fourth-power radiant heat transfer can be essentially proportional to temperature difference over a moderate temperature range, and conductive and convective coefficients can be similarly treated. The thermal capacitances also must not change with temperature. These qualifications allow the differential equations that follow to be of the linear type with constant coefficients.

In this analysis, the three conductors generate over time a constant heat loss, represented as $Q_o$. Prior to zero time, the conductors are de-energized, and the conductors and conduit are at ambient temperature. After zero time, the currents in the three conductors are equal and such that the heating generated in the conductors, $Q_o$, is constant. Also, the heat generated in the conductors consists of Joule heating, dielectric losses, and shield losses in the cables. Any small amount of heating in the conduit generated by induced currents is neglected. For the duration of the following derived responses, the ambient temperature is constant in this model.

The elements in Figure A.1 represent the following:

$Q_o$   Constant heat flow generated in unit length of the three cables (W)

$C_1$   Thermal capacitance of the three cables (W-h/°C)

$C_2$   Thermal capacitance of the conduit (W-h/°C)

$R_1$   Thermal resistance between cables and conduit (°C/W)

$R_2$   Thermal resistance between conduit and air (°C/W)

$q_1$   That time-varying portion of $Q_o$ stored in mass of the cables (W)

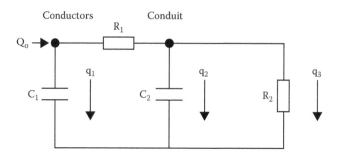

**FIGURE A.1**
Schematic representation of cable heat flow.

$q_2$  That portion of the time-varying heat flowing from the cables to conduit that is stored in the conduit's mass (W)

$q_3$  Time-varying heat flowing from the conduit to air (W)

In the equations that follow

ΔT  Steady-state conductor temperature rise above ambient reached after the transient responses (°C)

Φ  Time-varying temperature rise or difference (°C)

°C  Temperature degree difference of the Kelvin or Celsius scales

$\Phi_c$  Φ cables (°C)

$\Phi_d$  Φ conduit (°C)

$K_1$  Time constant equal to $R_1 C_1$ (h)

$K_2$  Time constant equal to $R_2 C_2$ (h)

$K_3$  Time constant equal to $R_2 C_1$ (h)

## Solutions

$$Q_o = q_1 + q_2 + q_3 \tag{A.1}$$

$$\frac{1}{C_2} \int_{t=0}^{t=T} q_2 \, dt = \Phi \text{ conduit, at time } t = T \tag{A.2}$$

$$\Phi_d = R_2 q_3 \tag{A.3}$$

$$\Phi_c = \frac{1}{C_1} \int_{t=0}^{t=T} q_1 dt = \Phi \text{ cables, at time } t = T \tag{A.4}$$

$$\Phi_c = \Phi_d + R_1(q_2 + q_3) \tag{A.5}$$

Taking the Laplace transforms of Equations A.1 through A.5

The Laplace transform of $Q_o$, a step change, is denoted by $\frac{1}{s}Q_0$.

The Laplace transform of $q_1$ is denoted by $\overline{q_1}$:

$$\frac{1}{s}Q_0 = \overline{q_1} + \overline{q_2} + \overline{q_3} \tag{A.6}$$

The Laplace transform of $\dfrac{1}{C_2} \displaystyle\int_{t=0}^{t=T} q_2 dt = \Phi_d = R_2 q_3$ is

$$\overline{\Phi_d} = \frac{1}{sC_2}\overline{q_2} = R_2\overline{q_3} \tag{A.7}$$

The Laplace transform of $\Phi_c$ is

$$\overline{\Phi_c} = \frac{1}{sC_1}\overline{q_1} \tag{A.8}$$

Solving for $\overline{q_2}$ using (A.6) through (A.8)

$$\overline{\Phi_c} = \frac{1}{sC_1}\overline{q_1} = \frac{1}{sC_2}\overline{q_2} + R_1\left(\overline{q_2} + \overline{q_3}\right)$$

$$\overline{q_3} = \frac{1}{sR_2C_2}\overline{q_2}$$

$$\frac{1}{sC_1}\overline{q_1} = \frac{1}{sC_2}\overline{q_2} + R_1\overline{q_2} + \frac{R_1}{sR_2C_2}\overline{q_2}$$

Multiplying by $sC_1$

$$\overline{q_1} = \left(\frac{C_1}{C_2}\overline{q_2} + sR_1C_1\overline{q_2} + \frac{R_1C_1}{R_2C_2}\overline{q_2}\right) \text{ and } \overline{q_1} = \left(\frac{C_1}{C_2} + sR_1C_1 + \frac{R_1C_1}{R_2C_2}\right)\overline{q_2} \tag{A.9}$$

Substituting in (A.6)

$$\frac{1}{s}Q_0 = \left(\frac{C_1}{C_2} + sR_1C_1 + \frac{R_1C_1}{R_2C_2}\right)\overline{q_2} + \overline{q_2} + \frac{1}{sR_2C_2}\overline{q_2}$$

$$\frac{1}{s}Q_0 = \left(\frac{C_1}{C_2} + sR_1C_1 + \frac{R_1C_1}{R_2C_2} + 1 + \frac{1}{sR_2C_2}\right)\overline{q_2}$$

$$Q_0 = \left[R_1C_1s^2 + \left(1 + \frac{C_1}{C_2} + \frac{R_1C_1}{R_2C_2}\right)s + \frac{1}{R_2C_2}\right]\overline{q_2} \qquad (A.10)$$

Taking the coefficient of s and changing some terms and adding fractions

$$\left(1 + \frac{C_1}{C_2} + \frac{R_1C_1}{R_2C_2}\right)s = \left(\frac{R_1C_1}{R_1C_1} + \frac{R_1C_1}{R_1C_2} + \frac{R_1C_1}{R_2C_2}\right)s$$

$$= \frac{R_1C_1(R_2C_2) + R_1C_1(R_2C_1) + R_1C_1(R_1C_1)}{R_1C_1R_2C_2}s$$

Substituting the expressions for $K_s$

$$\frac{K_1(K_2) + K_1(K_3) + K_1(K_1)}{K_1K_2}s$$

Rewriting (A.10)

$$Q_0 = K_1\left[s^2 + \frac{K_1 + K_2 + K_3}{K_1K_2}s + \frac{1}{K_1K_2}\right]\overline{q_2}$$

The quadratic coefficient $\left[s^2 + \dfrac{K_1 + K_2 + K_3}{K_1K_2}s + \dfrac{1}{K_1K_2}\right]$

has two roots expressed as $\left(s + \dfrac{1}{K'}\right)\left(s + \dfrac{1}{K''}\right)$.

These roots are real and one is larger than the other. Let $\dfrac{1}{K'}$ be the lesser root and $\dfrac{1}{K''}$ be the greater root. Then $K'$ is larger than $K''$:

$$\overline{q_2} = \frac{Q_0}{K_1} \frac{1}{\left(s + \dfrac{1}{K''}\right)\left(s + \dfrac{1}{K'}\right)} \qquad (A.11)$$

The solution to (A.11) can now be found as the antitransform of $\overline{q}_2$ (see Roberts and Kaufman, 1966, p. 189, number 77):

$$q_2 = \frac{Q_0}{K_1} \frac{1}{\frac{1}{K''} - \frac{1}{K'}} \left( e^{-\frac{t}{K'}} - e^{-\frac{t}{K''}} \right)$$

The expressions for the roots are found from the classical quadratic solution. The second term in the solution for the roots, which contains the radical, is

$$\pm 1/2 \sqrt{\frac{(K_1 + K_2 + K_3)^2}{K_1^2 K_2^2} - \frac{4}{K_1 K_2}}$$

Rewriting this term as

$$\frac{\pm 1}{2K_1 K_2} \sqrt{(K_1 + K_2 + K_3)^2 - 4K_1 K_2}$$

Defining $\rho \equiv \sqrt{(K_1 + K_2 + K_3)^2 - 4K_1 K_2}$ \hfill (A.12)

The expression for the roots becomes

$$\frac{1}{K''} = \frac{(K_1 + K_2 + K_3) + \rho}{2K_1 K_2} \quad \text{and} \quad \frac{1}{K'} = \frac{(K_1 + K_2 + K_3) - \rho}{2K_1 K_2}$$

$$K' = \frac{2K_1 K_2}{K_1 + K_2 + K_3 - \rho}; \hfill (A.13)$$

$$K'' = \frac{2K_1 K_2}{K_1 + K_2 + K_3 + \rho} \hfill (A.14)$$

$$\frac{1}{K''} - \frac{1}{K'} = \frac{\rho}{K_1 K_2}, \quad K'K'' = K_1 K_2, \quad K' - K'' = \rho \hfill (A.15)$$

$$q_2 = Q_0 \frac{K_2}{\rho} \left( e^{-\frac{t}{K'}} - e^{-\frac{t}{K''}} \right) \hfill (A.16)$$

$$\overline{q_3} = \frac{1}{sR_2C_2}\,\overline{q_2} = \frac{1}{sK_2}\,\overline{q_2}$$

$$q_3 = \frac{1}{K_2}\int_{t=0}^{t} q_2 dt = \frac{1}{K_2}\int_{t=0}^{t} Q_0\,\frac{K_2}{\rho}\left(e^{-\frac{t}{K'}} - e^{-\frac{t}{K''}}\right)dt \qquad (A.17)$$

$$\Phi_d = R_2\,q_3 \qquad (A.3)$$

$$\Phi_c = \Phi_d + R_1\left(q_2 + q_3\right) \qquad (A.5)$$

$$\Phi_c = R_1 q_2 + \left(R_1 + R_2\right)q_3$$

$$\Phi_c = Q_0 R_1 \frac{K_2}{\rho}\left(e^{-\frac{t}{K'}} - e^{-\frac{t}{K''}}\right) + \left(R_1 + R_2\right)\left[Q_0 - \frac{Q_0}{\rho}\left(K'e^{-\frac{t}{K'}} - K''e^{-\frac{t}{K''}}\right)\right]$$

$$K' - K'' = \rho \quad \text{and} \quad \Delta T = \left(R_1 + R_2\right)Q_0$$

$$\Phi_c = R_1 Q_0 \frac{K_2}{K' - K''}\left(e^{-\frac{t}{K'}} - e^{-\frac{t}{K''}}\right) + \Delta T - \frac{\Delta T}{K' - K''}\left(K'e^{-\frac{t}{K'}} - K''e^{-\frac{t}{K''}}\right)$$

$$\Phi_c = \Delta T\left[1 - \frac{1}{K' - K''}\left(K'e^{-\frac{t}{K'}} - K''e^{-\frac{t}{K''}}\right)\right] + R_1 Q_0 \frac{K_2}{K' - K''}\left(e^{-\frac{t}{K'}} - e^{-\frac{t}{K''}}\right)$$

$$(A.18a)$$

$$\Phi_c = \Delta T\left[1 - \frac{1}{K' - K''}\left(K'e^{-\frac{t}{K'}} - K''e^{-\frac{t}{K''}}\right) + \frac{1}{K' - K''}\frac{K_2 R_1}{R_1 + R_2}\left(e^{-\frac{t}{K'}} - e^{-\frac{t}{K''}}\right)\right]$$

$$(A.18b)$$

$$\Phi_c = \Delta T\left[1 - \frac{1}{K' - K''}\left(K' - \frac{K_2 R_1}{R_1 + R_2}\right)\right]e^{-\frac{t}{K'}} + \Delta T\left[\frac{1}{K' - K''}\left(K'' - \frac{K_2 R_1}{R_1 + R_2}\right)\right]e^{-\frac{t}{K''}}$$

$$(A.18c)$$

# Appendix B: Finding the Electrical Characteristics of the 400 Hp Motor

Motor analysis uses several motor equations as follows. Motor performance data given by manufacturers are based on operation at a standard voltage. The most common are 460 and 2300 V, phase to phase. Motor voltages allow a 4% voltage drop between the source voltage (480 and 2400 V) and the motor.

Shown in Figure B.1 is the changed circuit from Chapter 10 for one phase of the balanced three-phase motor. Rm and Xm represent the excitation's losses and reactance. Rm is very large and is neglected. Rs and Xs represent the stator resistance and leakage reactance. Xr represents the rotor leakage reactance referred to the primary side of the transformer. Rr represents the rotor resistance referred to the primary side of the transformer. Rr multiplied by $(1 - S)/S$ represents the load.

The most elemental equations are $(Hp)(0.746) = kW$ shaft output to the load's shaft.

$(Hp)(0.746)/(efficiency)(power factor) = kVA$ input

$(kVA input)(power factor) = kW$ input

The angle between the stator input voltage and stator current, Is, is $\varphi$.

Power factor $= \cos \varphi$

Reactive kvars $=$ kVA input $\sin \varphi$

Reactive kvars/kW input $= \tan \varphi$

In Bose (2002) and Mohan (2014), the following terms are derived. Using the equivalent circuit of Figure B.1

$$Is = Ir + Im$$

$S =$ slip rpm/synchronous rpm

$Zr = [Rt^2 + Xt^2]^{1/2} = Vs/Ir$

$Rt = Rs + Rr/S$

$Xt = Xs + Xr = (2\Pi$ Hz of applied voltage) $(Ls + Lr)$

$\psi$ sync $\equiv 2\Pi$ (synchronous rpm) $0.746/33,000$ kW/lb-ft

$\psi = \psi$ sync $(1 - S)$kW/lb-ft, which is the output power

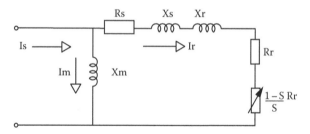

**FIGURE B.1**
The equivalent circuit used for analysis.

*This is a different definition than the ω in Chapters 2 and 10, wherein ω is revolutions per second. Here ʋ is used, in kW/lb-ft. In other texts, this symbol, ω, is used* for kW/lb-ft.

Synchronous rpm ≡ $rpm_{sync}$ = (applied-voltage frequency in Hz)(60 s/min). This is true for two-pole motors, where synchronous rpm is 3600 rpm. At 60 Hz

$$ʋ\,sync = 0.51134 \text{ kW/lb-ft @ 60 Hz}$$

Power output is $ʋ\,τ\,10^3$ = the mechanical power output in watts.

The sum of power output plus mechanical shaft losses is the mechanical power delivered across the air gap. Shaft losses are bearing losses plus cooling-air pumping losses.

The mechanical power delivered across the air gap is = $Ir^2\,[Rr\,(1 - S)/S]$.

$$\text{Furthermore, } (ʋτ\,10^3) + \text{(bearing losses and cooling-air losses)}$$
$$= (Vs^2/Zr^2)\,[Rr\,(1 - S)/S]$$

The mechanical power delivered across the air gap plus rotor copper loss is the total power delivered across the air gap, $Ir^2\,Rr/S$.

The term $I^2\,Rr\,(1 - S)/S$, the mechanical power delivered across the air gap, includes mainly the power output but also the small bearing losses and cooling air. Fans on the rotor inside the bearings circulate cooling air within the stator. The TEFC frame has outside axial fins, a shroud over them, and an outside fan forcing airflow through the shroud.

The mechanical power output in kW, $ʋ\,τ$, is defined as shaft power output to the load. To shaft power output is added the shaft mechanical power losses. These are small. The total shaft power is the output power plus the small mechanical losses. Because the shaft mechanical losses are small, they are neglected; thus

$$\text{Power output is } ʋ\,τ\,10^3 = Ir^2\,[Rr\,(1 - S)/S]$$

and, when finding total losses, are added back in as estimates.

The stator copper loss is $Is^2$ Rs; the rotor copper loss is $Ir^2$ Rr. Added to the stator copper loss is the excitation loss, which is small. The rotor copper loss, $Ir^2Rr$, is not included in the mechanical power output, but rotor resistance and load, Rr/S, are included in the impedance. The term $I^2Rr(1 - S)/S$ stands for the mechanical power delivered across the air gap. It excludes the rotor resistance. The term Rr/S includes the rotor resistance.

Breakdown is where the torque is maximum. Torque is limited by the motor's resistances and reactances, not by magnetic saturation. Motor analysis assumes that the flux is in its linear range with motor current. Slip at breakdown is

$$Sbd = Rr / \left[ Rs^2 + Xt^2 \right]^{\frac{1}{2}}$$

## The Analysis

Four steps are followed:

1. Find the values for the circuit resistances at full load.
2. Find the slip rpm, circuit values, and breakdown operation at overload conditions.
3. At 1300 lb-ft and 1800 rpm, find the current, voltage, and reactance.
4. With these characteristics, find the losses, power factor, and efficiency.

## Given

400 Hp motor, 460 V 60 Hz, 3570 rpm TEFC NEMA type.

Current at full load is 421 A, at 1/2 load is 220 A.

Torque at full-load is 588 lb-ft, at 1/2 load is 293 lb-ft.

Torque at overload is 1481 lb-ft; current is 1688 A; speed is 3466 rpm.

Rpm at full load is 3570, and 1/S is 120.

Power factor at full load is 92.6; sin $\Phi$ is the sine of the angle whose cosine is 0.926, or 0.3775.

Efficiency at full load is 96.2%.

Step 1. Impedance looking into the stator, phase volts divided by phase amperes, may be expressed in complex form by use of the power factor. Because the power factor is the cosine of the angular difference between the current and voltage, just as the impedance multiplied by the power factor is the real part, the impedance multiplied by the sine of the cosine's angle gives the reactive part. At full load, the power factor is listed as 92.6%. Thus, the sine is 37.75%.

a.  At full load, Zin = (460/1.732)/421 A = 0.63085:

$$0.63085(0.926) = 0.58417$$

$$0.63085(0.3775) = +j0.2381$$

$$Zin = 0.58417 + j0.2381 = 0.63085 \text{ at an angle of } 22.18°$$

b.  At full load, solving for rotor current, Ir: At no load, the current is 45 A. The iron losses are small; therefore, this current is almost all the winding reactive current, Im. This current is proportional to the voltage. In Figure B.1

$$Im = -j45 \text{ A}, \quad \text{and} \quad Ir = Is - Im$$

$$Is = 421 \text{ A at an angle of} - 22.18° = 389.8 - j158.9 \text{ A}$$

$$Ir = 389.8 - j158.9 - (-j45) = 389.8 - j113.9 = 406.1 \text{ A at an angle of} - 16.29°$$

$$(1-S)/S = 1/S - S/S = 1/S - 1: \text{At full load} 1/S = 120, \text{and} 1/S - 1 = 119$$

This excludes rotor copper losses. Estimating iron losses at 300 W and cooling-air losses and bearing losses as 200 W (see the following text).

At full load, Ir² (120 Rr) is the power across the air gap, which is the power-out with rotor copper losses:

$$Pout = 99.47 \text{ kW}$$

$$(407.5)^2 (119 \, Rr) = 99,400 + 200 = 19,760,694 \, Rr = 99.600 \text{ kW}$$

$$Rr = 99,600/19,760,694$$

$$Rr = 0.005040 \text{ ohms}$$

$$\text{Rotor copper loss} = (407.5)^2 (0.005040) = 837 \text{ W}$$

Rotor copper loss is the largest component of the rotor losses. Hence, the estimate of 200 W for the other losses: Total rotor losses: 1040 W

c.  KVAin = (460/1.732)(421/1000) = 111.82 kVA at *full load*.
    With power factor, kWin = 103.5 kWin.
    Losses based on power-in less power-out: At full load

$$kWin = 103.5 \text{ kW in per phase}$$

$$kWout = (400 \text{ Hp})(0.746)/3 = 99.47 \text{ kWout}$$

Difference is 4.03 kW total losses.
Efficiency at full load is 96.2%.
At full load: kWin = 103.5 kW, and (103.5 kW)(1 less efficiency) = (103.5 kW)(0.038) = 3.93 kW.
These values are based on given voltage, current, and power factor, with given torque and rpm. These two values, 4.03 and 3.93 kW, agree within ±1.3%.

$$kWout = (400 \text{ Hp})(0.746)/3 = 99.47$$

$$kWout = Ir^2 \, Rr(1/S-1) = (406.1)^2 \, Rr \, 119 \; 10^{-3} \text{ kWout}$$

$$Rt = Rs + 120 \, Rr = 0.62777 = Rs + 0.6084 : Rs = 0.62777 - 0.6084 = 0.0194$$

d.  Using 4.03 kW for full-load losses, rotor losses are 1.04 kW and stator losses are 2.99 kW. These stator losses are the winding copper loss and the iron loss. Most of this loss is the winding copper loss.
    Let iron loss equal 300 W and winding copper loss equal 2700 W.

$$Is^2 \, Rs = 2700 \, W : Is^2 = (421)^2, \quad \text{and} \quad Ir = 389.8 - j158.9 - (-j45)$$

$$= 389.8 - j113.9 \, A \quad Ir = 406.1 \, A \text{ at an angle of} - 16.29°$$

$$Rs = 2700/177241 = 0.0152 \text{ ohms}$$

Losses at full load:

| | |
|---|---|
| Stator copper loss | 2700 W |
| Stator iron loss | 300 W |
| Rotor copper loss | 840 W |
| Rotor bearing, iron, and air loss | 200 W |
| | 4040 W |

e.  Solving for Xr: Zr = Rr + jXr, and Xr/Rr = tan φr, Ir = 407.5 @ −16.96°. Hence φr = 16.96°.

$$Rr/S = 0.6048 \text{ ohms}, \quad \text{and} \quad \tan φr = \tan 16.96° = 0.30497$$

$$Xr = (0.6048)(0.30497) = 0.1845$$

$$Zr = 0.6048 + j0.1845 = 0.6323 \ @ \ 16.96°$$

$$Xs = 0.01876 \sin 22.16°$$

$$Rr = 0.00504 \ \text{ohms}$$

$$Rs = 0.0152 \ \text{ohms}$$

Step 2. At overload operation

Recalculate circuit elements for this region of operation.

KVAin = (265.6 V)(1688 A) = 448.33 kVA

Overload torque is 1481 lb-ft; current is 1688 A; speed is 3466 rpm.

Shaft power-out is (1481)(6.28)(3466)/33000 = 976.8 Hp.

976.8 Hp = 728.7 kW = 242.9 kW per phase output

kWout = 242.9 kW

Estimate shaft mechanical losses at 500 W.

kWair-gp = 242.9 + 0.5 = 243.4 kWout = 243,400 W = Ir² Rr (1 – S)/S

a.  Given slip rpm = 3600 – 3466 = 134 rpm, torque is 1481 lb-ft and torque times 2Π(3466)0.746/33,000.

$$kWair\text{-}gp = 242.9 + 0.5 = 243.4 \, kW.1/S = 3600/134$$

$$= 26.87 \, 1/S - 1 = 25.87$$

Find Ir for overload

Estimate angle of Ir from the following:

Estimate Ir as value of Is.

Ir = 1688A and Zr = 265.6/1688 = 0.1573 ohms

Ir² Rr (25.87) and Rr = (243.4) 10³/(1688 ²)(25.87) = 0.0033

Rr (1 – S/S) = (0.0033)(25.87) = 0.085

Rs = 0.02 estimate from above, making Rt = 0.105 ohms

Zr = 265.6/1688 = 0.1573 = 0.105 + jXt. And cos φ = 0.105/0.1573
= 0.669

*Estimate of φ is 48.03°.*

$$Is = 1688 \ A \ @ \ 48.03° = 1129 - j1255 \ A$$

$$Ir = Is - Im = 1129 - j1255 - (-j45) = 1129 - j1210$$

$$Ir = 1655 \ @ \ - 46.98°$$

b. Solving for the calculated circuit elements at 60 Hz overload operation

$$Rr = 243,400 \text{ W}/1655^2 (25.87) = 0.003435 \text{ ohms}$$

$$\text{and} \quad Rr/S = (0.003435)(26.87) = 0.0923$$

$$Zs = 265.6 \text{ V}/(1655 @ - 46.98° \text{ A}) = 0.1605 \text{ ohms} @ 46.98°$$

$$= 0.1095 + j0.1173 \text{ ohms}$$

$$Rs = 0.1095 - 0.0923 = 0.0172$$

$$Xt = j0.1173 \text{ ohms}$$

$$Rt = 0.1095 \text{ ohms}$$

$$Rr = 0.003435 \text{ ohms}$$

$$Rs = 0.0172 \text{ ohms}$$

c. Breakdown at 60 Hz

$$Sbd = Rr/\left[ Rs^2 + Xt^2 \right]^{\frac{1}{2}} = Sbd = 0.003435/\left[ 0.0172^2 + 0.1173^2 \right]^{\frac{1}{2}}$$

$$= 0.003435/0.11855$$

$$Sbd = 0.02898$$

$$Sbd = \text{slip rpm}/3600$$

$$\text{slip rpm} = (3600)(0.02898) = 104 \text{ rpm and shaft speed} = 3600 - 104 = 3496 \text{ rpm}$$

$$1/S = 34.51(1/S) - 1 = \left[ (1/0.02898) - 1 \right] = 33.511 - S = 0.971$$

$$\textit{w} \, \text{bd} \, \tau = [2\Pi(3,496) \, 0.746/33,000] \, \tau \text{ and } \textit{w} \, \text{bd} = 0.49657 \text{ kW/lb-ft}$$

$$Zbd = \left[ (Rr/S + Rs)^2 + Xt^2 \right]^{1/2}$$

$$Rt^2 = \left[ (0.003435) \, 34.51 + 0.0172 \right]^2 = 0.13574^2 = 0.018425$$

$$Xt^2 = 0.1173^2 = 0.013759$$

$$Zbd = (0.018425 + 0.013759)^{1/2} = 0.1794 @ 40.83°$$

$$Ibd = 265.6 \ V/0.1794 = 1480 \ A$$

$$Watts \ out = (1480)^2 \ Rr \ 33.51 = 252130 \ W$$

$$Watts \ out + mechanical \ losses \ of \ 500 \ W = 252,630 \ W = 252.63 \ kW$$

$$\tau \psi bd = 252.63 \ kW: \tau bd = 252.63 \ kW/0.49657 \ kW/lb\text{-}ft = 508.75 \ lb\text{-}ft$$

$$\tau motor = (508.75 \ lb\text{-}ft) \ 3 = 1526 \ lb\text{-}ft$$

Step 3. Operation at 1300 lb-ft and 1800 rpm

a.  At 30 Hz, synchronous speed is 1800 rpm. The motor is to operate at 1800 rpm so the applied frequency and synchronous speed
must be greater. There are two unknowns, slip rpm and rotor current Ir. With slip known, power-out can be found because torque
is known. Also, current can be found because S and Rr are known.
The breakdown value of slip rpm, calculated above at 60 Hz, is
about 100 rpm. At the new applied frequency, a little more than 30
Hz, the breakdown slip frequency will be about 100 rpm or less.
Hence, the applied frequency will be less than 32 Hz.

Estimate the applied frequency as 31 Hz. Then synchronous
rpm is 1860, with slip of 60 rpm. The desired phase torque is
1300 lb-ft/3, or 433.3 lb-ft. Estimate the torque to be this value
and solve for the power-out and current. If results are false,
assume another frequency higher or lower than 31 Hz, but 32 Hz
or less, and iterate the calculations.

b.  Calculate Ir at 1860 synchronous rpm and assumed torque of
433.3 lb-ft:

$$S = (60/1860) = 0.03226$$

$$1/S = 31.0$$

$$1/S - 1 = 30.0$$

$$1 - S = 0.9677$$

$$\psi \tau = [6.283(1860)(0.9677)(0.746)/33,000](433.3 \ lb\text{-}ft)$$

$$\psi \tau = 110.78 \ kW = 110,780 \ W$$

Adding other shaft losses that are estimated at 200 W, mechanical power developed is 110,980 W:

$$110,980 = Ir^2 \ Rr(1/S - 1) = Ir^2 (0.003435)(30.0) = 110,980 \ W$$

$$Ir^2(0.003435)(30.0) = 110,980\,W = Ir^2(0.10305)$$

$$Ir^2 = (110,980\,W/0.10305) = 1,076,953\,A^2$$

$$Ir = 1037.8\,A$$

Estimate Ir as 1040 A:

$$1040^2 = 1,081,600\,A^2$$

$$I^2\,Rr(1/S-1) = (1,081,600)(0.003435)(30.0) = 111,459\,W$$

$$= 111.5\,kW, and\,subtract\,200\,W$$

$$\ddot{w}\tau = (0.25566)\tau = 111.5\,kW$$

$$\tau = 111.5\,kW/0.25566 = 435.3\,lb\text{-}ft$$

Power-out is 111.3 kW/phase for a total of 333.9 kW, or 448 Hp. With estimates of 31.0 Hz driving frequency and Ir of 1040A, a torque of 435.3 lb-ft results. This is essentially the required operation. This illustrates the procedure for iterating assumed operating conditions to converge on the required conditions. Iteration starts with an assumed applied frequency and slip rpm.

c. Calculate applied voltage:

$$Xt = at\,60\,Hz, j(2\Pi60)(Ls+Lr) = 0.1173\,ohms$$

$$Xt = 0.1173\,ohms\,at\,60\,Hz$$

$$\ddot{w} = (2\Pi)(3,466)(0.746)/33,000 = 0.49205\ kW/lb\text{-}ft$$

At 31 Hz, $\ddot{w} = (2\Pi)(1,860)(0.746)/33,000 = 0.26419\ kW/lb\text{-}ft$

$$At\,31\,Hz, \quad Xt = 0.1173(31/60) = 0.06061$$

$$Xt = at\,31\,Hz, \quad j(2\Pi)(31)(Ls+Lr) = 0.06061\,ohms$$

$$Rr/S\,at\,31\,Hz = (0.003435)(31) = 0.10649$$

$$Rt_{31Hz} = Rr/S + Rs = 0.10649 + 0.0172 = 0.1237\,ohms$$

$$Zr_{31Hz} = 0.1237 + j0.06061\,ohms = 0.1378\,@\,26.1°\,ohms\,at\,31\,Hz$$

$$Ir = 1040(at\,minus\,angle\,of\,Zr = -26.1°) = 934.0 - j457.5\,A$$

$$Vs = Ir\ Zr = (1040)(0.1378) = 143.3 \text{ phase volts}$$

$$Vs = (143.3)1.732 = 248 \text{ V phase to phase}$$

Step 4. Calculate losses, power factor, and efficiency:
  a.   Find $Is = Ir + Im = Ir - j45 = 934.0 - j457.5 - j45 = 934.0 - j502.5$

$$Is = 1060.6 @ - 28.28°$$

$$Is^2\ Rs = (1,124,872)\ (0.0172 \text{ ohms}) = 19,348 \text{ W (see Figure B.1)}$$

$$Ir^2\ Rr = (1,081,600)(0.003435 \text{ ohms}) = 3,715.3 \text{ W}$$

| | |
|---|---|
| Stator copper | 19.35 kW |
| Stator iron | 0.50 kW |
| Rotor copper | 3.72 kW |
| Rotor mechanical | 0.20 kW |
| Total per phase | 23.77 kW |

Motor losses at required high-torque operation are 71.4 kW.
  b.   Calculate power factor and efficiency:

$$Is = 1060.6 @ - 28.28 = 934.0 - j502.5 : \varphi = 28.28°$$

$$\text{Power factor} = \cos\varphi = 0.881 = 88.1\%$$

$$Pout = 111.3\,kW\,(\text{from Step 3b mentioned earlier})$$

$$Eff = Pout/Pout + losses = 111.3/(111.3 + 23.8) = 82.4\%$$

## Results

Table B.1 compares the circuit-element values derived under full load and at overload and breakdown operations. They are not the same. Different simplifications in the model and data for the different operating modes show that each operating domain has different circuit-element values. Only the data for the overload and breakdown regions are valid for operation in the near-breakdown region.

**TABLE B.1**
Results of Characteristics and Values

| Hp | Is | Load | rpm | 1/S | Rr | Rs | Rt | Xt | PF % | Eff % | Losses |
|---|---|---|---|---|---|---|---|---|---|---|---|
| | | | | | At 60 Hz and 460 V | | | | | | |
| 400 | 421 | Full | 3570 | 120 | 0.005040 | 0.0194 | 0.6278 | 0.2033 | 96.2 | 92.6 | 4.04 kW |
| 977 | 1688 | Overload | 3466 | 26.87 | 0.003435 | 0.0172 | 0.1095 | 0.1173 | — | — | — |
| 1014 | 1480 | Breakdown | 3496 | 34.51 | 0.003435 | 0.0172 | 0.1357 | 0.1173 | — | — | — |
| | | | | | At 31 Hz and 248 V | | | | | | |
| 448 | 1061 | Required | 1800 | 31.0 | 0.003435 | 0.0172 | 0.1237 | 0.06061 | 88.1 | 82.4 | 23.8 kW |

*Note:* Except Hp, all values are for the equivalent circuit.

# Glossary

**ANSI**  American National Standards Institute.

**ASD**  Adjustable-speed drive. See VFD.

**BIL**  **Breakdown insulation level:** The insulation rating of a bushing's or a transformer's winding. Standard values are found in ANSI IEEE C57.

**Breakdown torque**  An induction motor's limit of torque that the motor can supply to a load coupled to the motor's shaft. The limit results from the motor's impedance elements, at less than magnetic saturation. Breakdown torque is about 220% of rated torque, a motor's thermal limit.

**CFM**  **Cubic feet per minute:** A measure of air flow.

**CT**  **Current transformer:** A donut-shaped laminated steel core with a spiral winding around it; a power conductor is installed through the donut hole and induces a current in the winding, which measures the cable's current.

**EPR**  **Ethylene propylene rubber:** A power cable insulation. Another insulation is XLP.

**ICEA**  Insulated Cable Engineers Assoc.

**IEEE**  Institute of Electrical and Electronic Engineers.

**IGBT**  **Integrated gate bipolar transistor:** This power transistor can be rapidly switched on and off to make a pulse-width modulated current waveform of different controlled amplitudes and frequencies.

**IN WC**  **Inches water column:** A measure of the static head of an operating fan or blower.

**kVA**  **Kilovolt amperes:** A load's power draw as the product of the applied kilovolts and load amperes that includes the phase angle difference between the voltage and current.

**kVAR**  **Reactive kilovolt amperes, referred to as kilovars:** A load's reactive power draw as the product of the applied kilovolts and load amperes multiplied by the sine of the power-factor angle. In industrial loads, currents lag the voltage in phase because the loads are dominated by induction motors, and induction motors require reactive, lagging excitation.

**kW**  **Kilowatts:** A load's power draw: the kilovolt amperes multiplied by the cosine of the phase angle difference between the voltage and current. The cosine is called the power factor.

**NEC**  National Electrical Code.

**NEMA**  National Electrical Manufacturers Association. This association maintains standards for most electrical equipment.

**OA**  Obsolescent for ONAN. The old ANSI designation before adoption of the European ONAN.

**ONAN**   Oil natural air natural. The code for the kVA rating of oil-filled transformers, which are equipped with internal oil circulating pumps and fans blowing on their cooling fins, when the pumps and fans are off. Then the oil circulates naturally and the fins cool naturally. With pumps and fans operating, the kVA rating is greater. The ONAN rating is the kVA reference for transformers' impedances.

**PLC**   **Programmable logic controller:** An electronic signaling equipment that accepts sensors' wiring and processes the signals into output terminals that connect to controls. A master digital signal connecting to a master controller is also part of the equipment, as well as a keyboard port and digital port for a display device, like a laptop.

**PT**   Potential transformer. See VT.

**Pull-out torque**   Same as breakdown torque.

**RTD**   Resistance temperature detector.

**RTU**   **Remote terminal unit:** A digital equipment connected into a distributed control network that ties local signals into the network. The unit can also have control capability for the local equipment generating and receiving signals. Related to PLC.

**TEFC**   **Totally enclosed fan cooled:** Designation for this type of motor in NEMA MG 1.

**THD**   **Total harmonic distortion:** The measure of a waveform's deviation from that of an alternating sine wave, caused by harmonics contained in the waveform.

**TXFR losses**   **Transformer losses:** The excitation losses and the windings losses. Excitation losses are the hysteresis losses. These are proportional to the applied voltage. Winding losses are the joule losses, which are proportional to the square of the current. At full load and rated voltage, losses are about 2%, with winding losses 1.6%. If percent efficiency is known, losses in percent are one minus efficiency. On air-cooled transformers, fan motor losses are included.

**UPS**   **Uninterruptable power supply:** A power-electronic equipment that converts a typical power source to a power supply that is without voltage sags, interruptions, spikes, and dips.

**VFD**   **Variable-frequency drive:** A power-electronic equipment that supplies a motor's current at a controlled frequency and motor load. It is provided with normal standard voltage and fixed-frequency power.

**VRLA**   **Valve-regulated lead acid:** A lead-acid battery made of sealed cells that have pressure-regulating valves. The cells are dry inside. The hydrogen and oxygen generated in the cell recombine, except for a tiny amount that builds up over time. The valve periodically vents in a puff the built-up gases.

**VT**   **Voltage transformer:** A transformer that connects to a voltage source and measures the voltage with a secondary low voltage.

**XLP**   **Cross-linked polyethylene:** A power cable insulation.

# Codes and Standards

IEC 61800-2, Adjustable Speed Electrical Power Drive Systems—Part 2: General Requirements

IEC 60896-2, Stationary Lead-Acid Batteries—General Requirements and Methods of Test—Part 2: Valve-Regulated Type

IEEE Standard 242, *Recommended Practice for Protection and Coordination of Industrial and Commercial Power Systems—Buff Book*, ANSI/IEEE Std. 242-86

IEEE Standard 484, IEEE Recommended Practice for Design and Installation of Vented Lead-Acid Batteries for Stationary Applications

IEEE Standard 485, IEEE Recommended Practice for Sizing Lead-Acid Batteries for Stationary Applications in Electrical Power Systems

IEEE Standard 519, Recommended Practices and Requirements for Harmonic Control

IEEE Standard 835, Standard Power Cable Ampacity Tables

IEEE Standard 944, IEEE Recommended Practice for the Application and Testing of Uninterruptible Power Supplies for Generating Stations

IEEE Standard 1184, IEEE Guide for the Selection and Sizing of Batteries for Uninterruptible Power Systems

IEEE Standard 1187, IEEE Recommended Practice for Design and Installation of Valve-Regulated Lead-Acid Storage Batteries for Stationary Applications

IEEE Standard 1189, IEEE Guide for Selection of Valve-Regulated Lead-Acid (VRLA) Batteries for Stationary Applications

IEEE/ICEA Standard S-135/P-46-426, Power Cable Ampacities

IEEE C37.20.2-2015, Standard for Metal-Clad Switchgear

ANSI/NEMA C50.41-2012, American National Standard for Polyphase Induction Motors for Power Generation Stations

IEEE C57.12.00-2000, Standard General Requirements for Liquid-Immersed Distribution, Power, and Regulating Transformers

NEMA MG-1-2011, Motors and Generators

NEMA Standard PE 1, Uninterruptible Power System

NEMA Standard PE 5, Utility Type Battery Chargers

NEMA ICS 7.1, Safety Standards for Construction and Guide for Selection, Installation and Operation of Adjustable-Speed Drive Systems

NFPA 70, *National Electrical Code*, 2002 Edition. National Fire Protection Assoc., Quincy, MA

NFPA 70, *National Electrical Code*, 2014 Edition. National Fire Protection Assoc., Quincy, MA

UL Standard 1778, Uninterruptible Power Supply Equipment

# Bibliography

Anderson, H., Technical data—A reference for the power industry, Hubbell Anderson Power Systems, AEC-41.

Bateman Manuscript Staff, *Table of Integral Transforms*, vol. 1, McGraw Hill Book Co., New York, 1954.

Bose, B. K., *Modern Power Electronics and AC Drives*, Prentice Hall PTR, Upper Saddle River, NJ, 2002.

Boyce, W. E. and DiPrima, R. C., *Elementary Differential Equations and Boundary Value Problems*, 2nd edn., John Wiley & Sons, New York, 1969.

British Standard 171, *Specification for Power Transformers*, British Standards Institution, London, U.K., 1978.

Bulletin EHB 90, *Engineering Data for Copper and Aluminum Conductor Electrical Cables*, The Okonite Company, Ramsey, NJ, 1990.

Central Station Engineers of the Westinghouse Electric Corp., *Electrical Transmission and Distribution Reference Book*, 4th edn., East Pittsburgh, PA, 1964 (first edition January 1942).

Editorial Staff, *Underground Systems Reference Book*, Edison Electric Institute, New York, 1957, Chapter 10.

Gardell, J. and Fredrickson, D., Motor bus transfer applications issues and considerations, J9 Working Group Report, IEEE Power System Relay Committee, May 2012.

Grainger, J. J. and Stevenson, W. D. Jr., *Power System Analysis*, McGraw-Hill, New York, 1994.

Henry, R. E., Response of power cables to fast transient loads, in *IEEE Ind. Appl. Soc.*, Seattle, WA, 2004, Cat. No. 04CH37569C.

Henry, R. E., Cable sizing for fast transient loads, in *IEEE 2003 I&CPS Conference*, Cat. No. 03CH37435C.

Henry, R. E., Fast transient loads of low-voltage power cables, in *Industrial and Commercial Power Systems 51st Tech Conference IEEE Industrial Applied Society*, Calgary, Alberta, Canada.

Houston Wire and Cable Co., *Houston Wire and Cable Catalog*, Houston, TX, 1987.

Klein, M., *The Power Makers*, Bloomsbury Press, New York, 2008.

Kron, G., *Equivalent Circuits of Electric Machinery*, John Wiley & Sons, New York, 1951.

Mohan, N., *Advanced Electric Drives*, John Wiley & Sons, Hoboken, NJ, 2014.

Mor Electric Heating Assoc. Inc., *Salamander Ceramic Infrared Emitters Technical Manual: Physical Properties of Materials*, Comstock Park, MI, p. 13.

NEMA MG-1 14.32 Section IV Part 30, Operation of AC motors from variable frequency or variable voltage power supplies, or both.

O Day, J., Fan Laws: What they are and examples on how to use them, ATS Advanced Thermal Solutions, Inc., Norwood, MA, May 21, 2010.

Roberts, G. E. and Kaufman, H., *Table of Laplace Transforms*, W. B. Saunders Co., Philadelphia, PA, 1966.

Walsh, G., The effects of reclosing on industrial plants, in *Proceedings of the American Power Conference*, Chicago, IL, March 1961.

Wiseman, R. J., An empirical method for determining transient temperatures of buried cable systems, *AIEE Transactions on Power Apparatus and Systems* 72(pt. III), 545–562, 1953.

Young, C. C. and Dunki-Jacobs, J., The concept of in-phase transfer applied to industrial systems serving essential service motors, *Transactions of the American Institute of Electrical Engineers, Part II: Applications and Industry*, 796, 508–518, 1961.

# *Index*